GEOLOGY UNDERFOOT

in
Yellowstone Country

MARC S. HENDRIX

2011
Mountain Press Publishing Company
Missoula, Montana

The Geology Underfoot series presents geology with a hands-on, get-out-of-your-car approach. A formal background in geology is not required for enjoyment.

is a registered trademark of Mountain Press Publishing Company.

Library of Congress Cataloging-in-Publication Data

Hendrix, Marc S., 1963-
  Geology underfoot in Yellowstone country / Marc Hendrix.
    p. cm. — (Geology underfoot)
  Includes bibliographical references and index.
  ISBN 978-0-87842-576-1 (pbk. : alk. paper)
  1. Geology—Yellowstone National Park Region—Guidebooks.
  2. Yellowstone National Park—Guidebooks. I. Title.
  QE79.H46 2011
  557.87'52—dc22

                                      2010051894

Printed by Mantec Production Company, Hong Kong

Mountain Press Publishing Company
PO Box 2399 • Missoula, Montana 59806
(406) 728–1900
www.mountain-press.com

To my family

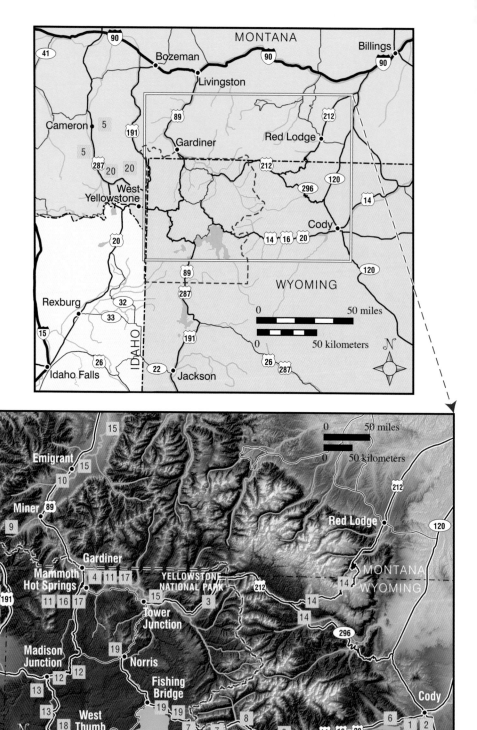

Sites featured in this book. Numbers in yellow squares
correspond to vignette numbers.

Contents

Preface

This book is designed to be an easy-to-understand introduction to the geology and geologic history of Yellowstone Country. What follows are short vignettes that highlight what I think are some of the main geologic features of this remarkable region. Some, such as Old Faithful, will be familiar to most Yellowstone visitors and deserve treatment because of their status. Most of the vignettes, however, are meant to acquaint the reader with aspects of the region's geology that may not be as well known but are equally, or perhaps even more, fascinating. Throughout the book I've tried to provide a sense of the tools and techniques scientists use to understand the geologic events preserved in rocks and sediments. I've also highlighted a few of the ways in which the world-class geology of Yellowstone Country has served, and will continue to serve, as a resource in the furthering of human knowledge. Above all, I hope the contents of this book will inspire readers to visit Yellowstone Country, take a few steps out of their vehicles, and see, with new eyes, the immensely powerful forces and wonderfully colorful history so well preserved in this one-of-a-kind international treasure.

The stops in the book occur in a variety of settings, including roadside pullouts, the shoulders of roads, and along trails maintained by the National Park Service and U.S. Forest Service. A few occur in the backcountry and require part or all of a day to access. At each stop, please exercise a high level of caution, particularly along Yellowstone's busy roads, on which drivers quite often are distracted by the scenery or various animals. Those readers willing to venture to stops off the road should bring drinking water, food, ample clothing, and sunscreen.

The high elevations of Yellowstone National Park often leave visitors surprisingly winded, and it is easy to underestimate the amount of exertion required to undertake what might seem like a relatively short stroll. Yellowstone's weather is also quite unpredictable. It is not unusual for seemingly mild sunny days to quickly evolve into something much less benign, like a summer cloudburst or a late spring or early autumn snow flurry. Remember that it is illegal to deface or take any rocks or other natural feature from the park, and the same applies to national forest land outside the park, where special permits are required for all collecting. As you enjoy Yellowstone Country, please treat it with the utmost care so future generations are able to experience its marvelous geologic features in the same way that you enjoyed them.

Acknowledgments

This book simply would not have been possible without the consistent support of my family, friends, and professional colleagues. In particular, I wish to acknowledge and thank James Lainsbury, the main editor with whom I worked at Mountain Press, for his continued encouragement and support, keen professional insight, and many hours of dedication to working on this project. I would also like to thank Beth Judy for the critical early role she played in getting me started with Mountain Press. I would like to thank my coworkers in the University of Montana Geosciences Department for numerous long and thoughtful conversations regarding many of the ideas presented in this book and my interpretation of the technical literature on those subjects. In this regard, I would especially like to thank Graham R. Thompson, who reviewed an early version of this manuscript, James W. Sears, and Steven D. Sheriff. Other colleagues I would like to acknowledge for the insights they provided regarding Yellowstone Country geology or other related topics addressed in this book include Steve Graham, Don Winston, Alan Carroll, Susan Vuke, Michael Hofmann, James Staub, Nancy Hinman, Jack Epstein, Judy Parrish, Lisa Morgan, Rob Thomas, Larry Smith, Mike Pope, Julie Baldwin, Joel Harper, Cheryl Jaworowski, Mike Stickney, and Rebecca Bendick.

I thank Aaron Deskins and Brian Collins for their help with the topography data used to construct the Getting There maps that appear at the beginning of each vignette. I am also grateful to Ron Blakey and the Northern Arizona University Geology Department for permission to use four paleogeographic reconstructions, and to Wade Johnson who graciously provided a high-resolution photograph of a hydro-thermal explosion. Many of the geologic interpretations presented in this book would not have been possible without the body of formal geologic literature that exists for Yellowstone Country, and several of the figures presented in this book were modified from figures published elsewhere. As such, I would like to formally recognize those works listed in the Sources of More Information section that appears at the end of the book.

This book required a dedicated field effort over parts of six field seasons and involved the assistance, support, and encouragement of many individuals. First among these is my father, Sherman S. Hendrix, who served as field assistant over four field seasons and provided the main inspiration for this work. Important additional field assistance for which I am grateful was provided by Brigette Hendrix, Matthew P. McArdle, Denison Von Maur, Greg Lovellette, and Charles Cash. Throughout this project, I received critical support and encouragement from my family and would like to thank especially my wife, Brigette, whose continuous support, encouragement, and companionship made this work possible, as well as my mother, Carol, and sister, Robin. Lastly, I wish to thank Gabriel and Michael Hendrix, our two sons, for providing important additional inspiration.

Introduction

Around the world, the name Yellowstone conjures images of stunning yet accessible natural beauty. Visitors to Yellowstone Country—the area within and surrounding Yellowstone National Park—are treated to magnificent scenes that include rugged mountains, colorful cliffs, and broad valleys often dotted with big game animals. Within the park itself, the rising steam and bizarre white landscapes of geologically active thermal features contrast with pastoral grassy meadows, some of which are strewn with large boulders. Those able to leave their vehicle and traverse on foot for even short distances can experience cool mists drifting upward from a wonderful array of waterfalls, sulfurous aromas billowing from cauldrons of bubbling hot mud, and geysers spewing scalding water skyward. Though much of the Yellowstone region includes sharp ridgelines and prominent peaks towering over deep, forested valleys, the heart of the park is a surprisingly subdued landscape where vast coniferous forests—many with sweeping scars still visible from wildfires—carpet a series of low, rolling ridges.

The sights and smells of Yellowstone are a direct result of the region's rich geologic history and the powerful tectonic forces that continue to shape it. In fact, Yellowstone National Park is centered over a single large volcano—the Yellowstone Volcano. Partially molten rock, called *magma*, exists as little as 2 miles (a little over 3 km) below the surface—an image that more than a few of the millions of annual visitors would probably find a bit startling! The heat associated with the volcano drives the many thermal features, including the numerous hot springs, mud pots, and geysers, for which the park is famous.

Why is the ground beneath Yellowstone Country so hot? To answer that, we need to first understand the structure of the Earth. It can be helpful to visualize Earth as an egg. The very outer layer of rock, the shell, is called the *crust*. The mantle, or egg white, underlies the crust and is about 1,780 miles (2,870 km) thick, extending to the core, or egg yolk, at the center of the Earth. The core has an outer layer made of liquid nickel and iron and a solid interior also made of nickel and iron.

By studying earthquake waves that pass through the mantle, like a sort of giant CAT scan, geophysicists have come to recognize that over long periods of time, hot pressurized rock of the mantle actually deforms like a slowly moving fluid—think of taffy. The deformation of mantle rock is not random but organized into currents. Heat generated from radioactivity in the mantle and outer core, along with residual heat left over from Earth's formation, causes rock at the base of the mantle to expand and rise, transporting heat from Earth's interior toward the surface. Near the top of the mantle this rock flows laterally and cools off, and then sinks back toward the core. Many geoscientists have suggested that mantle rock flows in roughly circular currents similar to those that form in a pot of soup on a hot stove.

A mass of hot, partially molten mantle is slowly rising in a channel-like current beneath Yellowstone from about 300 miles (500 km) below the surface. The hot rock creates a column of heat called a *thermal plume*, which is heating the overlying crust like a giant Bunsen burner and creating what geologists refer to as a *hot spot*. Hot spots are places where Earth's crust expands and partially melts. Typically, they are associated with volcanoes. Earth appears to have a few dozen hot spots. Besides Yellowstone, the island of Hawaii—renowned for its spectacular displays of hot magma spewing from the ground—is the manifestation of a thermal plume and its accompanying hot spot.

The hot spot currently associated with Yellowstone has been around for about 16.5 million years, although it has not always been located in the same place. Rather, since its inception, the hot spot has moved, relative to the overlying land, in a northeasterly direction about 1.5 inches (roughly 3.8 cm) per year, or about as fast as your fingernails grow. This slow drifting of the hot spot can be explained through the theory of plate tectonics. According to this widely accepted theory, the uppermost part of Earth's mantle and the overlying crust together compose a single rigid layer of rock called the *lithosphere*. The lithosphere is broken into about twenty large plates, like pieces of a puzzle that move very slowly relative to each other. The plates move over a hotter, more ductile layer of the mantle called the *asthenosphere*. Plates that include continents are between 12 and 43 miles (20 and 70 km) thick and made of granite in their upper part and a denser rock called *peridotite* in their lower part. In the oceans, the plates are between about 3 and 6 miles (5 and 10 km) thick and made mostly of dark volcanic basalt and peridotite.

All of North America is part of the North American Plate, which stretches from its eastern boundary in the middle of the Atlantic Ocean to the west coast. The North American Plate has been drifting to the

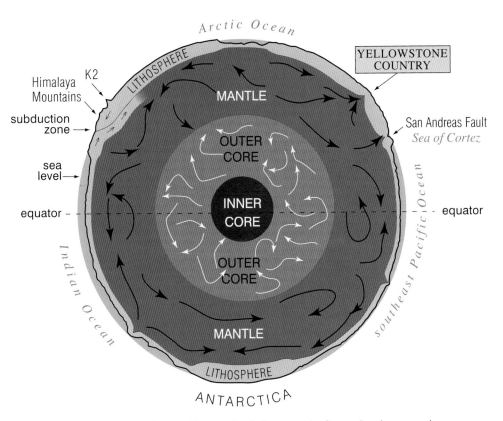

Cross section of Earth. Hot rock of the mantle flows in slow-moving currents *(arrows)*, which are shown here schematically. Geophysicists have discovered that a column of hot rock is slowly rising beneath Yellowstone, providing the heat that drives the volcanism and abundant thermal features for which the park is famous.

Looking north toward Middle and East buttes from US 26 about 40 miles (64 km) north of Pocatello, Idaho. At their core the prominent buttes consist of volcanic rock called *rhyolite* that intruded the crust after the Yellowstone hot spot passed through this area.

southwest over the thermal plume, which has remained in a relatively stationary position deep within the Earth. So through time, the hot spot has seemed to drift in the opposite direction—to the northeast—along the surface of North America, leaving behind volcanic evidence in parts of Oregon, Idaho, and Wyoming (see vignette 11 for more on the hot spot).

The boundaries between plates come in three flavors. One boundary type, called a *spreading center*, occurs where two plates are moving laterally away from each other. New lithosphere forms where the plates are diverging as magma wells up from the asthenosphere, cools, and hardens. A spreading center marks the eastern edge of the North American Plate in the middle of the Atlantic Ocean. A second boundary type develops where two plates are moving toward one another and one plate dives beneath the other, forming what's called a *subduction zone*. The diving plate, or the one that is being subducted, is made of oceanic lithosphere that is thinner and denser than the overlying plate, which can be either oceanic or continental lithosphere. As the diving plate reaches deeper and deeper into the mantle, it heats up and melts, producing magma that ascends to the surface of the overriding plate and erupts in volcanoes. Today, oceanic lithosphere of the Juan de Fuca Plate is diving eastward beneath the western boundary of the North

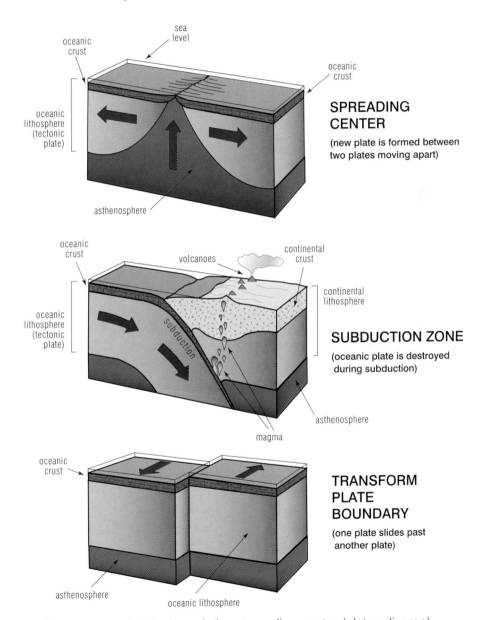

Three types of plate boundaries: spreading center (plates diverge), subduction zone (plates converge), and transform boundary (plates move past one another laterally). Arrows denote direction of movement.

American Plate, and ascending magma has produced the volcanoes of the Cascade Range in Washington, Oregon, and northern California. The third type of boundary, called a *transform plate boundary*, occurs where one plate slips horizontally past another. The San Andreas Fault in central and southern California marks this type of plate boundary.

The exact cause of plate movements is the subject of considerable debate. Many geoscientists have suggested that the lateral flow of mantle currents along the base of the lithosphere causes the plates to move. Another idea is that a plate slides downslope along its base, moving from a higher-elevation spreading center toward a lower-elevation subduction zone. The descent of cold and dense lithosphere into a subduction zone is thought to help pull the rest of the plate toward the subduction zone. Some geoscientists have even suggested that plate movements *cause* the mantle to flow, rather than the other way around. Regardless of the answer to this chicken-and-egg question, all geoscientists agree that lithospheric plates move laterally over the underlying asthenosphere. Although the details are obscured from our eyes by a vast amount of rock, tectonic activity within the asthenosphere and lithosphere has produced the Yellowstone hot spot and three cataclysmic volcanic explosions, each of which left behind a massive crater, called a *caldera*, in Yellowstone Country (see vignettes 11 and 12).

Although the thermal features related to the Yellowstone hot spot are arguably the region's most unusual geologic features, other aspects of the area's colorful geologic history can be easily seen and appreciated by park visitors armed with a little bit of observational know-how and a healthy imagination. For example, between 500 and 83 million years ago, prior to the heating of Yellowstone Country by the thermal plume, the area was inundated by ocean water at least six times (see vignettes 1–4). These slow advances and retreats left thick layers of sandstone and fossil-bearing limestone.

About 76 million years ago, due to changes in the subduction that was occurring along the west coast of North America, the crust was squeezed in an east-west direction, causing mountains to develop in the western portion of the Yellowstone region (see vignette 5). This mountain building passed eastward though Yellowstone about 60 million years ago, producing the initial topography associated with the northern Rocky Mountains (see vignette 6).

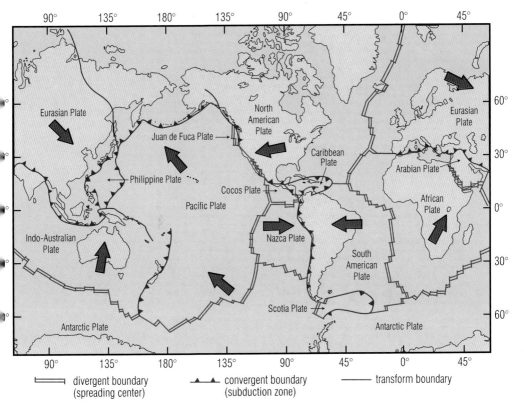

Map showing Earth's lithospheric plates and their direction of movement. Each plate moves slowly relative to the plates around it.

For reasons not well understood, around 50 million years ago a major phase of volcanism unrelated to the modern Yellowstone Volcano swept into the region. Volcanoes belched out tremendous volumes of ash, rock, and lava. A warm and humid climate promoted the growth of lush forests on the sides of the volcanoes and in the valleys between. The saturation of soil and ash by heavy rain occasionally caused sediment on the sides of the volcanoes to move downslope as powerful slurries called *debris flows*. These plowed through the forests, snapping off or uprooting trees and carrying them along. Many of the trees were petrified in the sedimentary deposits left by the debris flows (see vignettes 7–9).

The hot, humid environment began to grow cooler, and the volcanoes ceased to be active about 45 million years ago. Between this

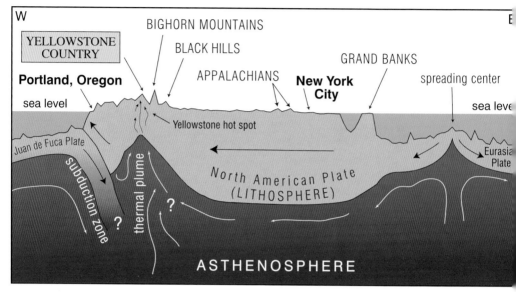

East-west cross section of the North American Plate, which extends from the spreading center in the middle of the Atlantic Ocean to the west coast of North America. The western plate boundary includes a subduction zone off the coast of the Pacific Northwest, a transform boundary that runs most of the length of California, and a spreading center that passes down the axis of the Gulf of California (Sea of Cortez). Partial melting of the Juan de Fuca Plate has produced the volcanoes of the Cascade Range in Oregon, Washington, and northern California. The North American Plate is drifting to the southwest relative to the thermal plume associated with the Yellowstone hot spot. The directions of flow (arrows) within the asthenosphere are generalized, but the upwelling of hot rock in the thermal plume is well documented.

time and 16 million years ago, little or no sediment accumulated in the Yellowstone region, producing in the rock record—what geologists "read" when they interpret the geologic history of an area—a physical surface called an **unconformity** (see vignette 1 for more on unconformities). Sediment began accumulating again between 16 and 14 million years ago and reflects the development of open grassland north of Yellowstone National Park. The grassland surrounded a lake basin out of which no water drained. The sediments that accumulated during this time have since hardened to sedimentary rocks that preserve a remarkable assemblage of fossil mammals, including an ancestral version of the modern horse (see vignette 10). The rocks also contain abundant volcanic ash that drifted into the area from eruptions south-west of the park. Some of the eruptions were those of the earliest volcanoes to develop as a result of the hot spot.

After forming, the hot spot drifted northeast about 2 inches (5 cm) per year and arrived in the Yellowstone region about 2.1 million years ago. Since then, three major eruptions have taken place, each causing the ground to collapse and form a caldera, which is Spanish for "cauldron." The oldest and largest caldera formed 2.1 million years ago, spans much of the park, and extends to the southwest beyond the park's boundaries (see vignette 11). A somewhat smaller caldera-forming eruption occurred about 1.3 million years ago southwest of the park. The most recent occurred 640,000 years ago and left a caldera that is centered in the park (see vignette 12). Although no caldera-forming eruptions have occurred since, dozens of smaller eruptions have mostly filled the depression of the youngest caldera with rhyolite (see vignette 13), creating the gentle topography associated with the central part of Yellowstone National Park.

About the time the hot spot arrived in Yellowstone Country, the first of several major ice ages began to grip the region. Most of the geologic evidence we have of this time period is related to the two most recent ice ages, which took place about 136,000 and 17,000 years ago. During each, glaciers in Yellowstone's higher-elevation regions flowed into the lower-elevation valleys and converged. As more snow accumulated, the volume of glacial ice grew, forming an ice-covered plateau, or ice cap, that stretched nearly across the entire park. Glacial ice flowed outward from the ice cap in all directions, spreading out under its own weight like a giant soufflé. Only the region's highest peaks stood above the ice cap's surface, which reached more than 11,000 feet (3,350 m) above sea level. The ice exerted a profound influence on Yellowstone's landscapes, carving much of the rugged relief of the region's mountainous terrain, leaving behind a variety of sedimentary deposits, and, when the ice melted, providing the water for several large floods (see vignettes 14–16).

As noted before, Yellowstone National Park is most famous for its numerous hot springs, geysers, and mud pots. In fact, Yellowstone contains the densest collection of thermal features on Earth, including more than two-thirds of Earth's geysers. Many of the thermal features occur within Yellowstone's youngest caldera, or north of the caldera margin between Norris Geyser Basin and the town of Gardiner, Montana. In this book, we'll examine limestone terraces at Mammoth (see vignette 17); geysers in Upper Geyser Basin, where Old Faithful is located (see vignette 18); and the craters left behind by hydrothermal explosions in Norris Geyser Basin and along the north shore of Yellowstone Lake (see vignette 19).

Along with the tectonic activity associated with the Yellowstone Volcano and the thermal plume below, the Yellowstone region is being pulled apart along with a larger portion of western North America that is undergoing extension. Known as the Basin and Range, this large physiographic province is so named because the extension has caused the crust to break, forming faults along which basins drop downward relative to intervening mountain ranges. The pulling apart of the crust has produced the rugged relief of the Madison Range west of Yellowstone National Park and caused two major faults to rupture in the region, leading to the 1959 Hebgen Lake earthquake and deadly Madison landslide (see vignette 20).

The incredibly rich and diverse geologic history of Yellowstone Country underscores the fact that geologic time is very deep. Time is to geologists like distance is to astronomers. Astronomers speak and think of distance in terms of light-years—the distance light travels in one year, moving at about 186,000 miles per second (299,340 km per second). Geologists think and speak of millions of years as representing relatively little geologic time. To organize the 4.6 billion years of Earth's history, geologists have developed a geologic timescale with formally recognized intervals of time. The first 90 percent of Earth's history, 4.6 billion to 542 million years ago, is classified as the Precambrian. Relatively little is known about the Precambrian because most of its rocks have been destroyed through erosion, and some of the techniques scientists use to study younger rocks simply don't work with Precambrian rocks. For example, scientists usually can't use fossils to gauge the age of Precambrian rocks because they rarely contain fossils. Most organisms preserved in the rock record evolved after Precambrian time.

The most recent 10 percent of geologic time is subdivided into three main eras. The Paleozoic Era (old time) lasted for about 291 million years, beginning around 542 million years ago and ending about 251 million years ago. This is the time frame during which the oldest sedimentary layers in Yellowstone Country were deposited, mostly when marine water inundated the region. The Mesozoic Era (middle time) lasted for 186 million years, from 251 until 65 million years ago, and is the time frame during which dinosaurs roamed much of the Rocky Mountain region. The most recent incursion of marine

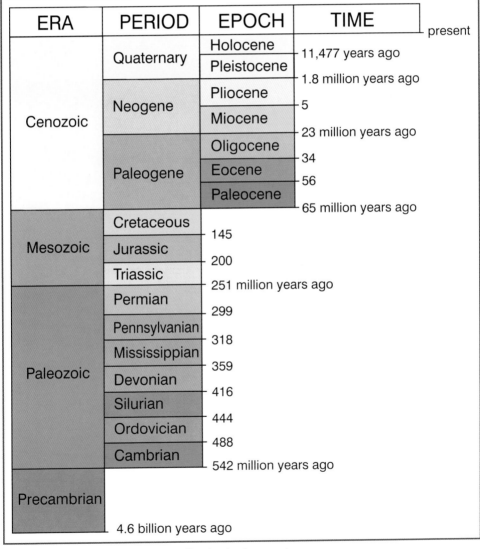

ERA	PERIOD	EPOCH	TIME
			present
Cenozoic	Quaternary	Holocene	11,477 years ago
		Pleistocene	1.8 million years ago
	Neogene	Pliocene	5
		Miocene	23 million years ago
	Paleogene	Oligocene	34
		Eocene	56
		Paleocene	65 million years ago
Mesozoic	Cretaceous		145
	Jurassic		200
	Triassic		251 million years ago
Paleozoic	Permian		299
	Pennsylvanian		318
	Mississippian		359
	Devonian		416
	Silurian		444
	Ordovician		488
	Cambrian		542 million years ago
Precambrian			4.6 billion years ago

Geologic timescale.

water into Yellowstone Country and the initial topographic expression of the Rocky Mountains both occurred during late Mesozoic time. The Cenozoic Era (new time), lasting from about 65 million years ago until the present, is the time during which most of the geological features in Yellowstone formed. The beginning of Basin and Range extension started during Cenozoic time, as did the inception of the Yellowstone hot spot and the carving of the region's glacial topography.

Comprehending the depth of geologic time is not easy. When compared to a standard 100-yard-long (90 m) American football field, with the goal line at one end representing the present, the Precambrian would take up almost 90 yards (80 m) of the field. The boundary between the Paleozoic and Mesozoic eras would be just outside the 5-yard (4.5 m) line, and the boundary between the Mesozoic and Cenozoic eras would be near the 1.5-yard (1.4 m) line. The most recent glaciation, which in the Yellowstone region peaked about 17,000 years ago, would be about 0.01 inch (0.25 mm) short of the goal line. The time that has elapsed since the Hebgen Lake earthquake would be about 0.0004 inch (0.01 mm) in front of the goal line—less than the average diameter of a single human hair!

The geologic timescale is continually in flux as dating techniques are developed and refined. Besides using the occurrence of fossils and various other techniques for determining the relative ages of rocks, geologists can actually assign a numeric age to some rocks by measuring the relative abundance of certain types of atoms within them. These atoms are radioactive and unstable. Over a period of time unique to each type of unstable atom, a tiny nuclear reaction occurs that results in the destruction of the unstable "parent" atom and the formation of an entirely different "daughter" atom. For example, uranium (parent) changes to lead (daughter), and potassium (parent) to argon (daughter). This process, called *radioactive decay*, begins as soon as an igneous rock has cooled below a certain temperature. By measuring, for example, the relative amounts of parent uranium and daughter lead in a given rock, keeping in mind that the rate of the creation of one and the destruction of the other is constant, geologists can determine the age of a rock.

A good example of how this technique has been used in Yellowstone involves the Yellowstone Volcano. Scientists have assigned ages to dozens of different lava flows and other igneous rocks erupted from the volcano. Using a potassium-argon technique, geologists determined that the most recent caldera-forming eruption occurred 2.053 million years ago, with an uncertainty of about 6,000 years.

Other technological advancements have helped geoscientists unravel mysteries of Yellowstone Country. Among these are sensitive earthquake-detecting devices called *seismometers*, which are used to determine the location of faults that rupture as well as the shape of the magma chamber under Yellowstone. Another useful modern tool is the network of sensitive Global Positioning System instruments deployed in and around Yellowstone. This network has provided much information about the deformation of the ground surface by the movement of magma and magmatic gases under the surface. Still another modern technology automatically monitors thermal water draining from Norris Geyser Basin. Every ten minutes, measurements of the amount and temperature of water passing over a small dam built across Tantalus Creek, located below the geyser basin, are sent to a satellite. The data provides scientists with information about the timing and magnitude of geyser eruptions in the basin and the amount of heat discharged by the basin's thermal waters.

Today, much of the ongoing monitoring of the Yellowstone region is conducted by the Yellowstone Volcano Observatory, a partnership between the University of Utah, U.S. Geological Survey, and National Park Service. Established in 2001, its main objective is to monitor the Yellowstone region for signs of renewed volcanism and related hazards and to improve the scientific understanding of the tectonic and magmatic processes that influence earthquakes, deformation of the ground surface, and thermal activity in the region. Given the very active nature of the geology within Yellowstone, the scientific results derived from this monitoring certainly will continue to provide a wealth of information about the unique international resource that is Yellowstone Country.

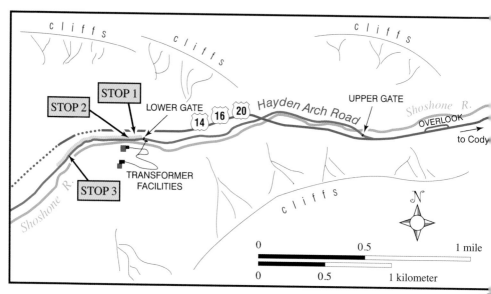

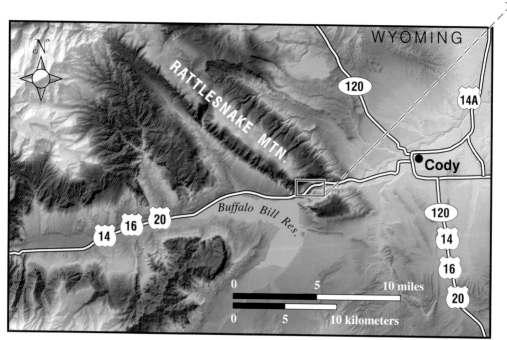

1.
The Missing Record of Deep Time
The Great Unconformity in Shoshone River Canyon

Geologic history is measured in time frames much bigger than those used to measure the history of humans. Geologists estimate that Earth is about 4.6 billion years old, about 900,000 times older than the earliest written human documents. Because it is so vast, geologic time is often referred to as *deep time*, and geologists are accustomed to thinking about geologic events in terms of millions or even billions of years.

In order to understand events that occurred over deep time, geologists often focus on places where different rock units meet, forming a boundary. One of the most significant types of these boundaries

---

GETTING THERE

From the rodeo arena at the western end of Cody, Wyoming, drive 1.8 miles (2.9 km) west on US 14/16/20 to Hayden Arch Road. Turn right (north) and proceed through the upper gate and over the Shoshone River. The upper gate is open Monday through Friday between 7 a.m. and 4 p.m. If you arrive after hours you'll have to make the traverse on foot or bike. Proceed past the upper gate about 1 mile (1.6 km) down the Shoshone River canyon to a lower gate, which is stop 1. If you are driving, park your vehicle at the small lot just before the gate. To reach stop 2, walk west along the road (downhill), past the gate for about 200 feet (60 m). You're looking for a spot at road level, on the north (right-hand) side, where layers of brown sandstone rest on top of grayish crystalline granite and gneiss. From stop 2, continue downhill about 0.25 mile (0.4 km) to stop 3, where you can examine steep cliffs of the granite and gneiss.

---

occurs where a layer of sedimentary rock rests directly on older igneous or metamorphic rock that formed deep beneath the surface. Such a boundary between younger sedimentary layers and older igneous or metamorphic rock is called an **unconformity** and is important because it represents the time it took for the igneous or metamorphic rock to be uplifted and eroded before the sedimentary rock was deposited on top of it.

By far the most geologically significant unconformity in Yellowstone Country is the contact between the region's oldest sedimentary rocks and the crystalline igneous and metamorphic rock that makes up the crust below. Because this crystalline rock extends deeply into Earth to the base of the crust and beyond, it is commonly referred to as **basement rock**. At the southern end of Rattlesnake Mountain in the Shoshone River canyon, west of Cody, Wyoming, Cambrian-age sandstone of the Flathead Formation rests directly on top of Precambrian-age basement rock. Because this particular unconformity represents about 1.5 billion years—roughly one-third—of Earth's history, it is sometimes called the Great Unconformity.

Geologists interpret Earth's history using the structures, composition, fossil content, and other features preserved in rocks of a known age. The cumulative body of evidence used to interpret conditions and events that occurred at different times during the geologic past is called the **rock record**. Unconformities are defined as both physical (space-related) and temporal (time-related) breaks in the rock record. In the temporal sense, the rocks below the unconformity may be much older than the rocks above the unconformity. The time that passed between these two ages is represented by the unconformity itself—no rock record exists for this time. In the physical sense, the unconformity is the contact or physical surface between younger layers of sedimentary rock and older rock. Because it is a boundary where the two rocks types meet and not a layer, the unconformity has no physical thickness.

At stop 1, look across the river to the small fenced parking lot with an electrical transformer station and a strange-looking cement structure that extends along the canyon wall before diving under the parking lot. The structure is a cement aqueduct containing pressurized water that spins a set of giant turbines below the parking lot, generating electricity. The cliff behind the transformer station is composed of gray granite and gneiss and pink pegmatite. This basement rock formed between 3.1 and 2.8 billion years ago and is the uppermost level of crystalline rocks that extend some 30 miles (48 km) down to the bottom of the crust. The vertical lines in the cliff mark drill holes into which dynamite was lowered to blast the canyon wall to make room for the transformer station. Their close spacing, 3 to 4 feet (0.9 to 1.2 m) apart, suggests that the rock presented a challenge to the engineers and construction workers who built the original transformer station at the beginning of the twentieth century.

If you look at the canyon wall exposed above the top of the transformer towers but below the horizontal section of cement aqueduct, you'll notice that the yellowish brown rock is clearly layered, or bedded, and the beds slope gently downstream (to the east, or left). This rock is sandstone of the Flathead Formation. The Flathead is the oldest rock of Cambrian age in northern Wyoming and adjacent states and represents the initial record of marine deposition here. In many locations on each of Earth's continents the sedimentary rock that rests directly on basement rock is Cambrian in age. This is no coincidence. Beginning around 540 million years ago, during the early part of Cambrian time, the world's oceans rose by as little as 320 feet (100 m) and as much as 820 feet (250 m), slowly flooding the continents and depositing sand, gravel, and mud. Vast areas that had been high and dry were covered by relatively shallow seas, which were much more hospitable to life than the windswept continents had been. In many places, Cambrian sedimentary rocks contain the oldest fossilized life-forms in a region, and the unconformity between the crystalline basement rock and these fossiliferous rocks marks the transition from a planet with a limited number of life-forms to one teeming with newly evolved organisms. (For more on one of these life-forms, see vignette 2.)

The Great Unconformity is where the Flathead sandstone meets the underlying gray gneiss and granite and pink pegmatite. By tracing individual beds of the lowermost sandstone from left to right across the exposure, you'll notice that each bed ends, or pinches out, against the unconformity, and that beds higher in the layered sequence terminate against the unconformity farther to the right. This subtle physical arrangement of sedimentary beds is called *onlap*. Onlap usually signifies that the rocks above an unconformity were deposited as water depth increased, information useful for reconstructing the geologic history of an area.

The onlap here suggests that the rising Cambrian sea slowly engulfed small hills of basement rock. As sea level rose and reached this part of Yellowstone Country, sand deposited in nearshore environments first filled in the low spots between the small islands of Precambrian rock. As sea level continued to rise, the encroaching layers of sand covered more and more of each island. Eventually, each island was completely buried. Through time, the sand hardened into the sandstone we see today.

After walking down the road 100 feet (30 m) or so from stop 1, you'll begin to notice several layers of brownish rock on the north side of the road that look suspiciously similar to the Flathead sandstone on the opposite side of the canyon. Indeed, visitors who get their nose to the outcrop will notice that the rock clearly consists of coarse sand. If you walk to the south side of the road and look down-canyon to the left, you'll see the same layers of sandstone sticking out of the canyon

Looking south across the Shoshone River from stop 1. The Great Unconformity *(dashed line)* occurs where the yellowish brown, layered Flathead sandstone onlaps the underlying metamorphic and igneous Precambrian rocks.

wall's north side. (Do not cross the guardrail for a better look because the ground overlooking the fast-flowing Shoshone River is very unstable.) The sandstone in the cliff contains great examples of *crossbeds*, cross sections of underwater sand dunes. The crossbeds are strong

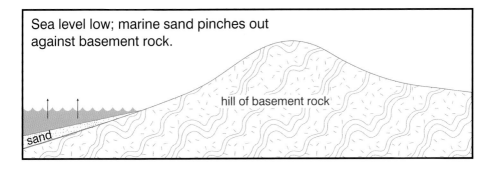

Sea level low; marine sand pinches out against basement rock.

hill of basement rock

sand

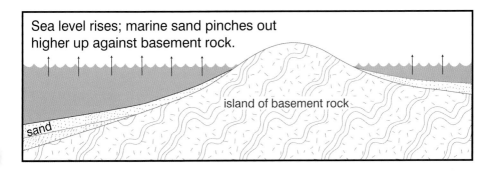

Sea level rises; marine sand pinches out higher up against basement rock.

island of basement rock

sand

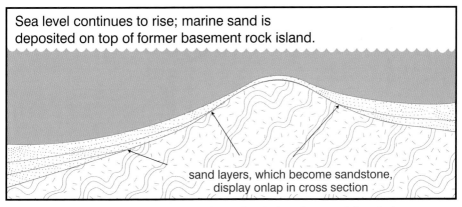

Sea level continues to rise; marine sand is deposited on top of former basement rock island.

sand layers, which become sandstone, display onlap in cross section

The world's oceans rose during Cambrian time and flooded the continents, depositing sediment, typically sand. In Yellowstone Country marine water deposited the sand that later hardened to form the Flathead Formation, which commonly onlaps the underlying Precambrian basement rock.

evidence that the sands were deposited in an energetic, shallow-water environment—a shoreline region capable of moving the sand around.

As you continue walking down the road, keep an eye on the rocks exposed in the outcrop to your right. Eventually, you'll come to the spot where the yellowish brown sandstone rests on top of unlayered, grayish granite and gneiss at road level. Here, at stop 2, you can put your hand on the Great Unconformity.

Individual grains of Flathead sandstone are well rounded. The rounding occurred as the sand grains were rolled around in a high-energy, shallow-water environment.

An outcrop of Flathead sandstone exposed in a cliff above the Shoshone River. The highlighted crossbeds were produced by vigorous shoreline currents in the Cambrian sea.

Perhaps the most impressive feature of the unconformity is what is missing: about 1.5 billion years of Earth history. The gneiss, granite, and pegmatite that make up the basement rock below the unconformity all formed deep within Earth's crust, probably at least 10 miles (16 km) below the surface. Therefore, before these rocks could be exposed at the surface, a substantial thickness of overburden had to be stripped away. This style of top-down erosion is called *erosional unroofing*. Although it may seem unlikely that miles of rock were removed to bring these deep-seated rocks to the surface, the substantial age difference between the basement rock and the Flathead sandstone would have been more than enough time for this erosion to be accomplished. In other words, the unconformity does not represent 1.5 billion years in which nothing happened, it represents a very long period of slow weathering and erosion. Around 540 million years ago, the granitic rock was exposed at the Earth's surface only to be reburied by sand.

The author's father, Sherman S. Hendrix, standing on the Great Unconformity *(dashed line)*. The unconformity represents about 1.5 billion years of geologic time for which no record remains in this area.

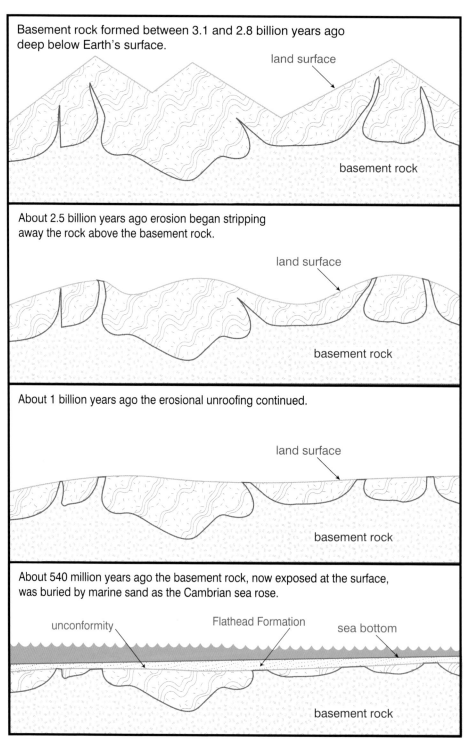

Basement rock formed between 3.1 and 2.8 billion years ago deep below Earth's surface.

land surface

basement rock

About 2.5 billion years ago erosion began stripping away the rock above the basement rock.

land surface

basement rock

About 1 billion years ago the erosional unroofing continued.

land surface

basement rock

About 540 million years ago the basement rock, now exposed at the surface, was buried by marine sand as the Cambrian sea rose.

unconformity

Flathead Formation

sea bottom

basement rock

The development of the Great Unconformity in Yellowstone Country.

After returning from your cerebral trip through deep time, continue down the road to stop 3 for views of steep cliffs of gray gneiss and granite that are sometimes adorned with rock climbers. Sheetlike dikes of pink pegmatite cut every which way across this basement rock. Well-formed horizontal dikes of pegmatite, up to 3 feet (1 m) thick, occur about halfway up the cliff. A few of the big blocks of crystalline basement rock near the road contain dikes of pink pegmatite that are about 1 foot (30 cm) thick. All of the dikes formed when water-rich magma within the Earth intruded fractures in the gneiss and granite. The gray gneiss is older than the granite, having been metamorphosed before being intruded by the granite. Because the dikes of pink pegmatite cut across the gneiss and granite, we know that it is the youngest of the three rock types.

Pegmatite is an igneous rock that has the composition of granite but contains unusually large crystals. In some pegmatite, single mineral crystals can be more than 10 feet (3 m) long, but those in the Shoshone River canyon aren't nearly that big. The pegmatite here contains large

A pink pegmatite dike cuts across older gray granite in the Shoshone River canyon.

pink crystals of potassium feldspar, many of which are more than 1 inch (2.5 cm) across. If the sun is out, look in the cliffs and blocks of rock by the road for shiny flat surfaces in the pegmatite where potassium feldspar crystals have broken along what is called a *cleavage plane*.

The atoms making up the crystals of potassium feldspar—or any mineral for that matter—have an orderly atomic arrangement. In situations where minerals grow freely in open space, the orderly arrangement of atoms is expressed as the sort of beautiful crystals that are sold in rock shops and commonly worn as neck pendants. In most cases, however, minerals grow in confined spaces, such as deep beneath Earth's surface, and do not form well-shaped crystals. However, even when crystals form in confined environments, the atomic arrangement that defines them still exists. Within the arrangement of atoms making up the crystals of feldspar are preexisting directions of weakness, or cleavage planes. Look for pinkish crystals of potassium feldspar and white crystals of plagioclase, another feldspar mineral. Both types exhibit good cleavage. Cleavage planes are similar to the grain of wood. And just as wood generally splits parallel to its grain rather than across it, feldspar crystals usually break along cleavage planes, resulting in the flat, parallel surfaces that reflect sunlight.

So why are the feldspar and other mineral crystals so big in pegmatites compared with those of "normal" granites, in which crystal sizes typically are smaller and more uniform? Granites start as molten rock called *magma*. As the magma cools, different minerals slowly crystallize within it. The minerals that crystallize first do not incorporate much water or other fluids, such as carbon dioxide, into their atomic structure. (At the high pressures that exist in a magma chamber, water and carbon dioxide are fluids that are dissolved in the liquid magma.) By the time most of the magma has formed solid rock, liquid magma that remains contains a high concentration of fluids, including water and carbon dioxide. In such residual magma, it is difficult for new mineral crystals to start growing because the atoms and molecules needed to form them are in a magma diluted by fluid. At the same time, the crystals that do form grow very quickly because the dissolved fluids carry the remaining atoms and molecules to the site of crystal growth. As a result, pegmatites usually have a small number of exceptionally large crystals, and they commonly occur as dikes because the fluids were injected into cracks that had formed in the solidified granite.

As you walk back to your car, away from the Precambrian basement rock, imagine yourself on the Cambrian beach that existed here: A vast sea extends to the west and north, and a hilly, windswept continent reaches to the east. Over the course of a few million years or less, the sea slowly rises and inundates the beach with seawater. Waves create strong currents that move the sand back and forth, shaping it into underwater dunes with internal crossbedding. As the sea rises, the sand is spread around by currents and eventually covers nearly the entire area. In vignette 2, we take a look at limestone that showcases evidence of life-forms and other depositional environments that existed in the Cambrian sea. The limestone is just up the road.

A single crystal of potassium feldspar in Precambrian pegmatite. The crystal broke along a cleavage plane, which reflects sunlight.

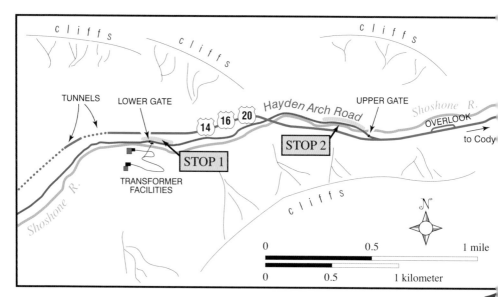

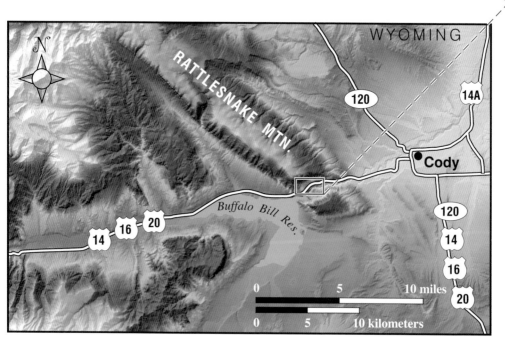

2.

Invasion of the Trilobites

Cambrian Sea Life Flourishes near Cody

As discussed in vignette 1, about 500 million years ago, during Cambrian time, marine water covered Yellowstone Country and deposited sandstone of the Flathead Formation. At that time the Yellowstone region was located along the sloping western edge of the North American continent. Although covered by ocean water, this area was still technically part of the continent because it was underlain by continental crust, mostly granite, and not oceanic crust, which is mostly basalt. Much of the North American continent was a broad submerged shelf with a centrally located bulge called the Transcontinental Arch. Even

---

GETTING THERE

From the rodeo arena at the western end of Cody, Wyoming, drive 1.8 miles (2.9 km) west on US 14/16/20 to Hayden Arch Road. Turn right (north) and proceed through the upper gate and over the Shoshone River to the lower gate—the starting point for vignette 1. The upper gate is open Monday through Friday between 7 a.m. and 4 p.m. If you arrive after hours, you'll have to make the traverse to the lower gate on foot or bike. From the lower gate, walk to the east back up the road about 150 feet (45 m) to stop 1, a layer of grayish green to brown rock on the north side of the road. To reach stop 2, proceed by vehicle or on foot back toward the upper gate. The outcrop you're looking for is about 150 yards (135 m) shy of the northwest end of the bridge. If you drive, it's best to park at the upper gate.

---

though sea level rose during Cambrian time, the bulge remained above sea level, quite unlike the continent's configuration today.

So how do geologists know all of this? To answer this question, it's necessary to review the Cambrian sedimentary record. Although Cambrian rocks were deposited across North America, they crop out widely in only two areas: in a belt in the Midwest, and down the length of the Appalachian Mountains. Within the Rocky Mountains, Cambrian sedimentary rocks are exposed only where they have been uplifted by tectonic activity. By determining the ages of exposed and subsurface rock (using rock cores or bits of rock from drilled oil wells), geologists have determined that deposition during Cambrian time started earlier on the margins of the continents and later in the continental interiors. West of the Transcontinental Arch, younger and younger beds of Cambrian rock pinch out farther and farther to the east against basement rock below them, forming a classic example of onlap (onlap is described in more detail in vignette 1). The onlap of Cambrian sediments onto North American basement rock reflects the rise in Cambrian sea level and the progressive flooding of the continent.

But there is other evidence that the sea grew deeper. Before heading to stop 1, you may want to take a moment to look at the brownish Flathead sandstone exposed at the parking area near the lower gate. The well-rounded sand grains and abundant crossbedding of the sandstone

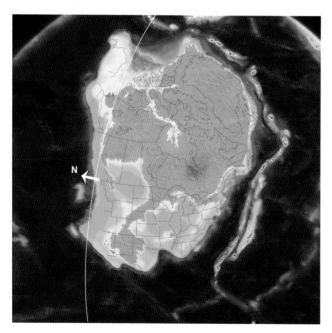

Reconstruction of North America about 500 million years ago. Most of the eastern and western United States was submerged, whereas the middle of the continent remained above sea level along what is called the Transcontinental Arch. The arch stretched across the midcontinental United States and into what is now southern Canada. —Courtesy of Ronald Blakey and the Northern Arizona University Geology Department

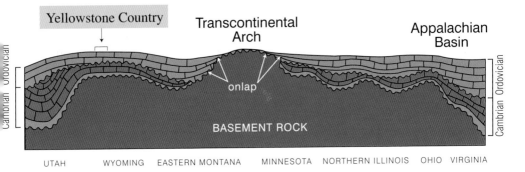

In North America, Cambrian and Ordovician sedimentary rocks, nearly all of which are marine, are thickest at the western and eastern ends of the continent. Because the sedimentary rocks onlap against the crystalline basement rocks below, geologists infer that the sediments were deposited as marine water slowly flooded the interior of the continent.

suggest it was deposited in shallow, high-energy conditions, which are characteristic of a shoreline environment, with crashing waves, rip currents, and other fast-moving water (see vignette 1 for more on the Flathead sandstone). From the parking area, proceed about 150 feet (45 m) up the road to stop 1, a layer of grayish green to brown rock on the north side of the road. This relatively thin rock unit, which doesn't form as prominent an outcrop as the sandstone, belongs to the Gros Ventre Formation. It rests directly on top of the Flathead sandstone and is a few million years younger. Unlike the relatively pure sandstone of the Flathead, the Gros Ventre contains abundant mud.

What brought about the change from sandy deposits to muddier deposits? Most geologists interpret this sequence of rocks as representing a deepening of the sea, otherwise known as a *transgression*, in which deeper water environments replaced shallow, high-energy environments as the shoreline migrated landward. When the sea first arrived in the region, it mostly deposited sediment derived from the weathering and breakdown of local exposures of Precambrian basement rock. Because this rock consists of granite, gneiss, and pegmatite, the sediment it produced was mostly quartz with some feldspar, as seen in the Flathead sandstone. However, as Cambrian time progressed and the basement rock became progressively buried, it was eliminated as a sediment source. The mud in the Gros Ventre Formation consists of clay and silt derived from rock that was still exposed above sea level farther east. Offshore currents carried the fine sediment to the Yellowstone region. Sand also occurs in the Gros Ventre Formation and was probably reworked from the underlying sand beds by bottom currents.

The Cambrian transgression did not happen quickly. It took millions of years for the shoreline to move from Yellowstone Country to the continental interior. In the relatively calm water of the deeper sea, small particles of mud and fine sand were able to settle out of the water onto the ocean floor. The dark color of the sandy shale of the Gros Ventre reflects the presence of abundant organic matter. Because land plants had not yet evolved by Cambrian time, much of the organic matter was probably derived from algae that lived in the sea.

The sandy shale of the Gros Ventre Formation has a mottled appearance, which reflects the presence of fossilized burrows in the rock. The

The author's father is standing next to an outcrop of the Gros Ventre Formation, a sandy shale deposited when the Cambrian sea was at least 10 feet (3 m) deep. The fine-grained character of the rock indicates it was deposited in deeper water than the underlying, relatively mud-free, coarser-grained Flathead sandstone.

burrows were produced by ancient organisms, many of which were probably deposit feeders—simple creatures who cruised the Cambrian seafloor, ingesting and extracting nutrients from mud. Trilobites—one type of deposit feeder—probably produced some of the burrows. This now-extinct arthropod was enjoying its heyday during Cambrian time.

Once you've had a chance to check out the Gros Ventre Formation, proceed to stop 2, a massive outcrop on the left (north) that towers over the road and the Shoshone River. These rocks consist of fine-grained, light-brown limestone of the Cambrian Gallatin Formation, the unit that overlies the Gros Ventre Formation. Limestone is a sedimentary rock composed mostly of calcite and a few other minerals that are based on carbonate—a molecule made of one carbon atom surrounded by three oxygen atoms. Limestone often forms in warm, shallow, tropical or semitropical marine water that doesn't have mud washed into it from nearby landmasses (see vignette 3 for more on limestone formation).

Marine organisms produced the churned appearance in this exposure of the Gros Ventre Formation as they burrowed into the sediment when it was at the bottom of the Cambrian sea. Quarter for scale.

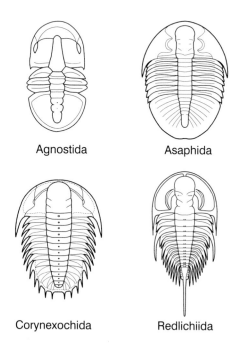

Agnostida   Asaphida

Corynexochida   Redlichiida

Trilobites were an amazingly diverse group of arthropods that evolved quickly and spread widely across the world's ocean bottoms during Cambrian time. They went extinct about 250 million years ago.

During Cambrian time, the equator ran nearly through the Yellowstone region from southwest to northeast, resulting in these conditions.

The limestone is composed of beds, each generally 3 to 4 inches (7 to 10 cm) thick, which represent deposition that occurred in a particular set of environmental conditions. When these conditions changed, deposition stopped temporarily, forming the visible upper surface of each bed. Some of the beds are composed of a jumble of flat, rounded pebbles. Called a *flat-pebble conglomerate*, this rock formed when sea level was low enough that the muddy seafloor partially dried out and cracked. When seawater flooded the region again, it ripped up chunks of the exposed mudflat, rounding them and producing flat pebbles. The pebbles were then redeposited with finer-grained sediment and later hardened as the flat-pebble conglomerate beds.

There are well-developed burrows in some of the light-gray limestone beds in the Gallatin Formation. Unlike the generally churned appearance of the Gros Ventre Formation, the burrows in the Gallatin Formation are more distinct, with many well-defined individual burrows. No one knows for sure what organisms produced these burrows, although they are similar to those produced by certain kinds of shrimp. It is not uncommon to see Cambrian-age sedimentary rock containing an abundance of burrows, as we see in the Gros Ventre and Gallatin formations. The burrowing reflects the thriving nature of life-forms that evolved during Cambrian time in environments ideal for their development.

Prior to the explosion of new marine life-forms during Cambrian time, the world's oceans and shorelines were dominated by simple forms of algae. The first simple algae in the fossil record, found in western Australia, are 3.5 billion years old. Between this first appearance of algae and the first appearance of trilobites during Cambrian time, at least 2 billion years passed during which there were very few animals eating the algae. Bits and pieces of it were deposited along the seafloor with other sediment, creating an organic-rich mud. As we saw with the dark-colored sandy shale of the Gros Ventre Formation at stop 1, at least some of the Cambrian mud deposited in Yellowstone Country was mud of this type. It would have been dark, stinky, and uninviting to us, but for a deposit feeder, such as a trilobite, it was home sweet home. The abundance of organic-rich mud and apparent lack of predation allowed the number of trilobite species and individuals to grow to the point where they became one of the dominant groups of

A flat-pebble conglomerate bed of limestone in the Gallatin Formation. The roundness of the pebbles indicates they were rolled around on the bottom of the shallow sea prior to being deposited. Quarter for scale.

organisms in the oceans, and thus the planet. It wasn't a glorious living, being a trilobite, but the conditions were such that they flourished.

According to Darwin's theory of evolution by natural selection, trilobites able to best take advantage of the environmental conditions that characterized the Cambrian seafloor out-competed trilobites that were not as well suited. However, with the exception of the Transcontinental Arch, the Cambrian seafloor stretched across North America, so there were many environmental nooks and crannies where trilobites of different sizes, shapes, and specific needs could evolve. In addition, it's quite likely that sea level rose and fell many times. The flat-pebble conglomerate here at stop 2 is strong evidence of this fluctuation. This episodic rise and fall changed the localized environmental conditions on the seafloor every few million years or so.

When a rise in the Cambrian sea flooded new parts of Earth's continents or deepened local waters to an ideal level for colonization, trilobites migrated en masse and evolved into new species that exploited the new or changed environments. Often more than one trilobite species evolved from a common ancestor, each filling a unique habitat in a pattern that scientists call *adaptive radiation*. Eventually, sea level would fall again and the favorable environmental conditions

The tubelike structures are innumerable burrows crisscrossing parts of the Gallatin Formation. They are evidence that marine organisms colonized the sea bottom. Rock hammer for scale.

would vanish, causing many trilobite species to become extinct. The drops in sea level were usually small, but they changed the water temperature and reduced the amount of oxygen available for respiration on the seafloor. Some habitats dried out altogether. This pattern of migration, radiation, and extinction, which scientists call *iterative evolution*, is characteristic of the Cambrian fossil record. Although geologists and paleontologists might argue over what specific environmental conditions changed to produce the migration, radiation, and extinction of different trilobite species, the evolution of trilobites during Cambrian time stands as one of the biggest and best-documented evolutionary expansions of a single group of organisms in the history of Earth.

Due to their abundance and wide geographical distribution, trilobites are one of the chief fossil types used to determine the specific age of rocks of Cambrian time. Trilobite fossil assemblages—collections of different trilobite species found together—tell paleontologists whether the rocks in which they occur were deposited during early, middle, or late Cambrian time. In fact, Cambrian time is subdivided into even shorter time intervals, called *stages*, which lasted a few million years and are defined by the presence or absence of certain trilobite species. Although trilobites provide the basis for the subdivision of the Cambrian period, not all Cambrian rock contains trilobites, so geologists aren't always able to assign a specific stage to every outcrop of it. Nevertheless, as a general rule, trilobites provide very useful Cambrian-age information for geologists.

As you examine the exposures of Cambrian rocks along the road, imagine yourself snorkeling around the seafloor, watching trilobites and other marine organisms crawl and eat their way across the ocean bottom. Not only did fluctuating sea level change what was deposited on the seafloor, from the organic-rich sandy shale of the Gros Ventre Formation to the flat-pebble conglomerates and heavily burrowed limestones of the Gallatin Formation, it gave rise to the trilobite, one of the most prolific groups of organisms in Earth's history. Although the Cambrian sea rose and fell long before Yellowstone's famous thermal features formed, these early events are, nonetheless, an important chapter in the history of deep time in Yellowstone Country.

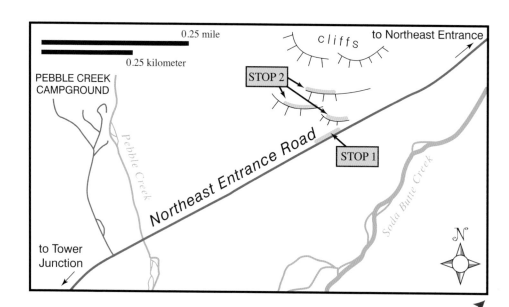

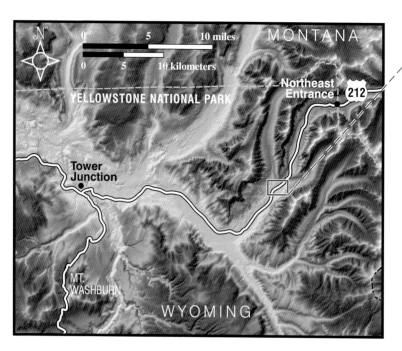

3.
Limy Record of Shallow Seas
Mississippian Limestone at Pebble Creek Campground

In vignettes 1 and 2 we examined the evidence that marine waters inundated Yellowstone Country and much of the North American continent during Cambrian time. This was not the only time period during which such flooding occurred. In fact, geologists recognize that there have been six major transgressions, or rises in sea level, during the last 500 million years, all of which temporarily inundated the entire region in and around Yellowstone.

One of the first geologists to recognize these transgressions was Northwestern University professor Larry Sloss. By comparing the distribution and thickness of the North American rocks deposited by these marine waters with those of the same age on other continents, particularly Asia, Sloss recognized that global sea level rose high enough six times to inundate much of the interior of each continent. Because the thick sequences of marine sedimentary rocks occur in the continental

GETTING THERE
Stop 1, a roadcut in Madison limestone, is located on the north side of Northeast Entrance Road in Yellowstone National Park. Drive to Pebble Creek Campground, which is 9.2 miles (14.8 km) southwest of Northeast Entrance and about 18.9 miles (30.4 km) east of Tower Junction. Park at the campground's sign-in spot and walk east along Northeast Entrance Road for about 0.4 mile (0.6 km) to the unassuming patch of light-gray limestone. To get to stop 2, a natural outcrop of the limestone featuring "tear-pants weathering" and "ball-bearing talus," proceed carefully uphill (north) for 50 to 100 yards (45 to 90 m).

interiors, far from the nearest oceans of today, he concluded that the rises must have been significant—possibly 330 feet (100 m) above modern-day sea level. He called the sedimentary deposits left by each transgression a *sequence* and named each after a North American Indian tribe. For example, the Cambrian sequence is named after the Sauk Indians, a tribe that lived around the yellow Cambrian-age rocks common in Michigan and Wisconsin. In addition, there are the Tippecanoe, Kaskaskia, Absaroka, Zuni, and Tejas sequences.

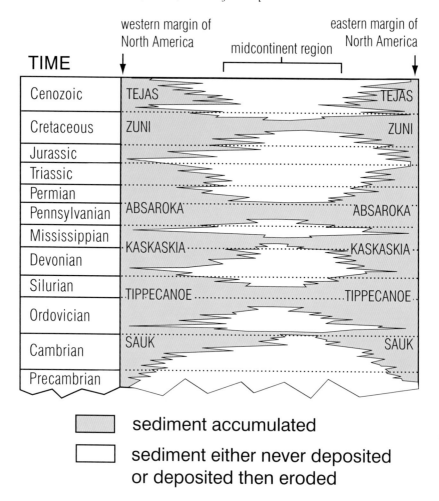

Simplified sketch of Larry Sloss's sequences and their distribution across North America. Six major transgressions of marine water inundated the North American continent after the end of Precambrian time.

Sloss included rocks of Mississippian age (359 to 318 million years ago) in the Kaskaskia sequence. Rocks of this sequence are represented in the Yellowstone area by the Madison limestone, which is about 1,000 feet (300 m) thick and was deposited about 350 million years ago at the bottom of a vast limy sea that stretched across much of the western part of the North American continent. The limestone crops out at the surface along a belt that stretches from Rattlesnake Mountain, just west of Cody, Wyoming, north and west across the northeastern corner of the park, and up the west flank of the Absaroka Range toward Bozeman, Montana. Within the park itself the Madison limestone does not crop out much, but it does occur widely in the subsurface and probably is the source for much of the limestone (called *travertine*) that is forming today in Yellowstone's thermal features (see vignette 17).

The roadcut at stop 1 is the only one in the park where Madison limestone is exposed. The best exposure is on the north side of the road. Angular blocks of gray Madison limestone litter the base of the roadcut. If you examine one, chances are you'll see a fossil—perhaps more than one. Many of the blocks are chock-full of fossils, forming what geologists refer to loosely as "fossil hash"—a rock consisting almost exclusively of fossils that were partially broken down by waves before being entombed in rock. A magnifying glass or hand lens will help you see the fossils since most of them are pea-sized or smaller.

The remains of crinoids are very common in the limestone. These ancient flowerlike organisms are related to the modern starfish, although most modern crinoid species float along in the water rather than live on the bottom. During Mississippian time, most were anchored to the seafloor with a rootlike structure called a *holdfast*. A long stem composed of many small segments of calcite extended from the holdfast to the creature's feeding apparatus, called a *calyx*, which resembled a flower. The "petals" strained particles of food from the surrounding water. When the crinoids died, they fell apart, and the calyx and little segments making up their stems rolled around like small pebbles on the seafloor. Eventually, these fragments were encased in lime mud or cemented together by calcite that precipitated between the individual fossil pieces, forming limestone.

You might also see brachiopods (a type of bivalve) and bryozoans (a branching or meshlike colonial organism) in the limestone. Some parts of the roadcut are composed of very clean, white lime sand, indicating that at times vigorous currents swept the sediment around, carrying fine-grained particles away and leaving behind the sand. Other parts contain a few fossils encased in a finer-grained matrix, suggesting quieter water conditions in which white limy mud could settle out onto the seafloor.

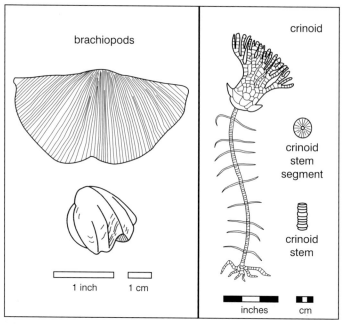

Line drawings of a crinoid and two types of brachiopod. Because they were segmented, crinoids broke apart when they died. Their stem segments are very common at stop 1.

The abundance of fossils in this roadcut is quite striking. In fact, these rocks are among the most fossiliferous in Yellowstone Country. But why? As with many other geologic mysteries, geologists look to modern geologic environments to understand those of the past. Today, abundant limestone is forming in shallow marine environments near the equator, where the water is warm and, in places, clear blue with very little suspended clay or silt in it. Between about 350 and 330 million years ago, the North American continent was situated in such a way that Yellowstone Country was located very close to the equator.

brachiopods

crinoid stem
segment

A typical chunk of the fossiliferous Madison limestone near Pebble
Creek Campground. The large curved lines are cross sections
through brachiopods, a type of bivalve that was very common during
Mississippian time. The smaller circular fossils are stem segments
from crinoids. Quarter for scale.

Warm seas covered the region, which was a broad, shallow continental
shelf. Although a modest series of islands, about 300 miles (480 km)
to the west, and part of the mid-continent, about the same distance
to the southeast, were both above sea level, these landmasses were
far enough away that most of the fine-grained silt and clay derived
from their weathering settled out of the sea before reaching the Yellowstone region. The warmth and clearness of the water provided ideal
conditions for the crinoids, bryozoans, and brachiopods that are now
fossilized in the Madison limestone.

As with the Gallatin limestone we examined in vignette 2, the
Madison limestone consists almost entirely of minerals that are based
on carbonate—molecules consisting of one carbon atom surrounded
by three oxygen atoms. By far the most common carbonate mineral in
the Madison limestone is calcite. Along with one or two other carbonate minerals, calcite precipitated out of the Mississippian sea through

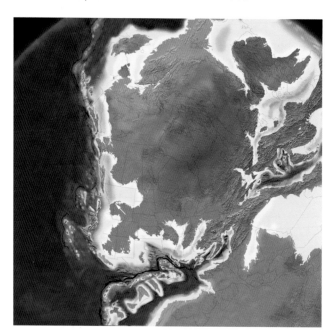

Reconstruction of North America 325 million years ago during Mississippian time. The Yellowstone region *(arrow)* was located on a shallow marine shelf about 300 miles (480 km) from the nearest exposed land. The clarity of the water and the fact that the equator ran nearby promoted the formation of limestone. —Courtesy of Ronald Blakey and the Northern Arizona Geology Department

various chemical reactions, becoming lime mud or sand. These carbonate-producing chemical reactions take place more readily in warm water than in cold water, so the fact that the equator passed nearly through the Yellowstone region resulted in favorable conditions. In addition, many organisms that are now fossilized in the Madison limestone secreted carbonate minerals to form their hard parts. These organisms were mostly filter feeders that used small, tentacle-like strainers to funnel plankton into their mouths. If there had been an abundance of suspended silt and clay in the water from weathering landmasses, the mud would have clogged the food-filtering systems of these organisms, reducing their abundance or causing them to simply die off. So the clearness of the Mississippian water favored their existence, as well.

The Madison limestone at stop 2, just above (north of) the roadcut at stop 1, is exposed in a natural outcrop. The rock is partially covered by loose gravel. Geologists call these marble-sized bits of limestone "ball-bearing talus" because they can send you for a ride if you walk across an outcrop covered with the stuff. To make matters worse, the sharpness of the weathered limestone can cut you deeply, although the damage is usually limited to clothing. Geologists say that limestone like this has "tear-pants weathering." So please be careful as you make your way around the outcrop looking for fossils. You don't want to slip on the ball-bearing talus and tear your pants—or worse!

The tear-pants weathering forms due to the manner in which limestone breaks down during the weathering process. Unlike granite or sandstone, which disaggregate, or break apart crystal by crystal or grain by grain, limestone like the Madison dissolves molecule by molecule. When slightly acidic rainwater falls on an outcrop, it collects in preexisting depressions on the rock surface. Over centuries and millennia, this process accentuates small variations in the rock surface as the water dissolves the low spots faster. Impurities in limestone, such as quartz sand or chert (finely crystalline quartz), tend to stand out in positive relief because they dissolve much more slowly than the carbonate minerals. The sharp edges characteristic of tear-pants weathering are the end result of this slow dissolution.

As you scour the outcrop for Mississippian fossils, or nurse wounds you received from slipping on the ball-bearing talus and experience firsthand the tear-pants weathering, pause for a moment to think about how different Yellowstone Country looked during Mississippian time. The region was a vast, flat area covered by a sea that stretched eastward to the Midwest, northwest to Alaska, and southward into Mexico.

The outcrop of Madison limestone at stop 2, north of Northeast Entrance Road.

The water would have been a deep blue color because of the lack of suspended clay or silt, and the underwater visibility would have been excellent. A scuba diver would have seen bryozoans swaying back and forth on the seafloor in the gentle currents and small fish darting among delicate crinoids.

Near the end of Mississippian time, around 325 million years ago, the clear blue sea that stretched across Yellowstone Country retreated. For approximately the next 275 million years, the Yellowstone region was mostly dry and flat, with relatively little sediment being deposited. Shallow seas would occasionally drift in and deposit thin layers of sediment. However, never again were such thick sequences of limestone deposited in the Yellowstone region as during Mississippian time.

The very rough surfaces that form on limestone outcrops in the arid West are sometimes called tear-pants weathering. Quarter for scale.

4.

The Cretaceous Interior Seaway
Marine Deposits in Gardner River Canyon

The Yellowstone region was most recently inundated by marine water during Cretaceous time. Called the Cretaceous Interior Seaway, this body of water formed as southward-advancing water from the Arctic Ocean met northward-advancing water from the Gulf of Mexico. The seaway reached Yellowstone Country around 120 million years ago and receded to the east around 83 million years ago, so for about 37 million years the Yellowstone region was underwater. At its maximum size, the seaway stretched from present-day New Orleans to Barrow, Alaska, covering almost the entire midcontinent.

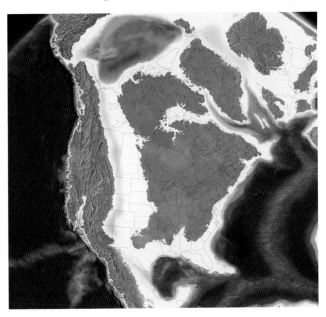

The relatively shallow Cretaceous Interior Seaway connected the Gulf of Mexico with the Arctic Ocean, flooding much of the North American midcontinent during Cretaceous time. A chain of mountains existed in what is today the Pacific Northwest.
—Courtesy of Ronald Blakey and the Northern Arizona University Geology Department

Rivers on the eastern side of the seaway flowed westward to its eastern shore, while rivers draining a series of mountains along the western side of the continent flowed eastward to the seaway's western shoreline. Sediment deposited by these rivers formed broad coastal plains that sloped gently toward both shorelines. The rivers were substantial in size, with swampy floodplains colonized by lush plant growth. Following the initial establishment of the interior seaway, marine water covered the entire Yellowstone region, and the western coastal plain was located west of Yellowstone in parts of southwestern Montana and Idaho. As the seaway slowly drained, the western shoreline migrated eastward through the Yellowstone region, and by the end of Cretaceous time, Yellowstone was located on the coastal plain.

Two major simultaneous geologic events caused the midcontinent to flood during Cretaceous time. First, extreme global warming completely or nearly melted both of Earth's polar ice caps, raising sea level about 250 feet (75 m). What evidence indicates to geologists that this global warming occurred? By measuring certain chemicals in the shells of fossil foraminifera (small planktonic organisms) that lived in the mid-latitudes of the Atlantic Ocean during Cretaceous time, scientists have concluded that peak sea-surface temperatures there were about 98 degrees Fahrenheit (37 degrees C), about 12 degrees Fahrenheit (7 degrees C) warmer than today. Paleontologists have found a wide variety of dinosaur fossils in the arctic regions of Alaska and plant fossils of temperate forests in Antarctica, indicating that warm temperatures allowed life-forms to flourish in places we wouldn't expect.

GETTING THERE

A spectacular display of sedimentary rocks deposited in the Cretaceous Interior Seaway is exposed at stop 1. This small roadside pullout on the west side of North Entrance Road is 1.6 miles (2.6 km) south of the official North Entrance of Yellowstone National Park. The pullout is 3.1 miles (5 km) north of the Albright Visitor Center and Museum at Mammoth Hot Springs. Stop 2 provides a great long-distance view of these sedimentary rocks. To get there from the Mammoth Hot Springs Hotel, drive south for 2 miles (3.2 km) on Grand Loop Road and turn right (west) onto Upper Terrace Drive (32.2 miles, or 52 km, north of Madison Junction). Proceed about 0.2 mile (0.3 km) and park in the small parking lot on either side of the road.

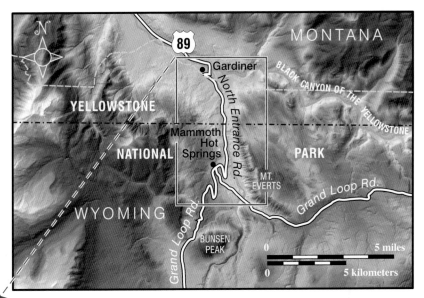

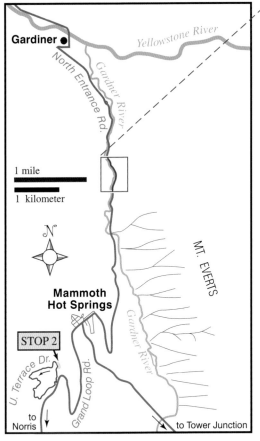

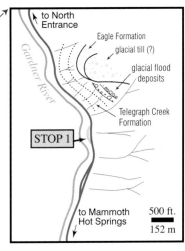

Were such a sea level rise to occur today, the world's major coastal cities and large parts of each continent would be mostly underwater, including much of Washington DC and Los Angeles, all of London and Tokyo, and nearly the entire Florida peninsula. Viewed in this context, the prospect of continued melting of today's polar ice caps is a sobering thought indeed.

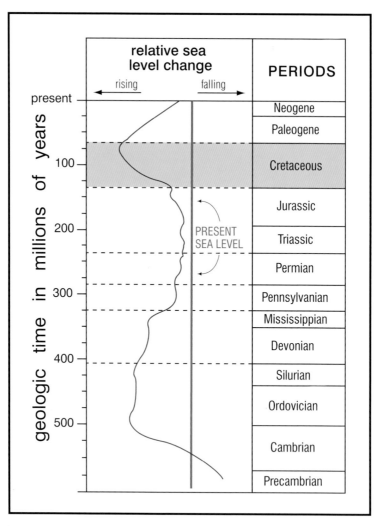

History of global sea level fluctuations. The detailed shape of this curve—exactly how fast and how much global sea level rose and fell through time—has been the source of much scientific controversy since the curve was published in the late 1970s, but the scientific community widely accepts that sea level rose during Cretaceous time.

The second event that led to the establishment of the Cretaceous Interior Seaway was the downward warping of the midcontinent. Geologic evidence suggests that a subduction zone existed along the western margin of North America beginning about 250 million years ago. Oceanic lithosphere of the Farallon Plate was being subducted eastward beneath the North American Plate, forming a volcanic arc along the continent's margin. According to one popular theory, by Early Cretaceous time, the rate of subduction increased, causing the western part of the overriding North American Plate to be compressed. Giant slabs of the upper crust were thrust over one another like a series of giant shingles, forming a mountain belt east of the volcanic arc but west of Yellowstone Country. This mountain belt, called a *fold and thrust belt* because of the compressional deformation that occurred within it, stretched from northern Mexico to southern Canada. The shingled slabs of rock within the belt thickened the crust and acted like a large weight, causing the lithosphere in front of the mountain belt to warp downward. It is this warped region that filled with seawater and became the Cretaceous Interior Seaway.

The sedimentary deposits of the seaway and Cretaceous coastal plain rivers are widespread across western North America and once blanketed Yellowstone Country. Although these rocks have since been partly eroded and then covered by younger sedimentary and volcanic rocks, they crop out widely across the northern part of Yellowstone National Park. The Gardner River canyon is an excellent place to view them.

Stop 1 is a small roadside pullout just south of Eagle Nest Rock, a great wall of gray and brown pinnacles that tower over the road where the canyon narrows. This area was named for golden eagles and other raptors that prefer the isolation the pinnacles provide for nesting and the open country nearby for hunting. Please be careful at this roadside pullout; the road is quite narrow and has relatively fast-moving traffic. I recommend against crossing the road to get a closer look at the outcrops. The cliffs are very unstable, and large boulders frequently fall down and accumulate on the talus slope at their base. Besides, the National Park Service has closed this area because it is a favored haunt of bighorn sheep.

If you would like to see the Cretaceous sedimentary rocks up a bit closer, proceed about 175 yards (160 m) north along the road to a small

East-west cross section through western North America during Late Cretaceous time. An increase in the rate at which the Farallon Plate was being subducted beneath the North American Plate produced tectonic compression, which caused a fold and thrust belt to form. The weight of this mountain belt caused the lithosphere in front of it to flex downward. Rising seas flooded this region and formed the Cretaceous Interior Seaway. Arrows denote direction of movement.

pullout on the west (left) side. From the pullout, a great overview of the sedimentary rocks can be had to the east, and, by carefully crossing the road, it is possible to examine fragments of the rock that have weathered off the cliffs. Please do not attempt to climb the cliffs—they are extremely unstable and dangerous.

Spectacular cliffs forming Eagle Nest Rock are well displayed to the north and east of stop 1. The upper half of the canyon wall consists of a brownish layer of glacial till containing a number of prominent vertical cracks. Several layers of very coarse boulders are exposed on the cliff underneath the till. The ancestral Gardner River deposited the boulders when it existed at a higher elevation, having been temporarily blocked by an earthflow (see vignette 16 for more on earthflows). The river eventually overtopped the earthflow dam and reestablished its course, depositing the boulders high on the canyon wall and cutting the steep narrow canyon that exists today.

The obviously layered gray rock closest to the road is the Telegraph Creek Formation, deposited in the Cretaceous Interior Seaway some 100 million years ago. The formation is composed of alternating layers of yellowish brown sandstone and darker gray shale. The base of the cliff is dominated by shale and has only a few thin layers of sandstone. These sandstone layers become thicker and are more numerous and tightly spaced toward the top of the cliff. At the north (left) end of the exposure, the very top of the cliff is composed entirely of sandstone belonging to the younger Eagle Formation.

The shale-dominated Telegraph Creek Formation formed from mud that accumulated in less-energetic offshore portions of the sea-way, where water depths were sufficient to prevent surface waves from churning the bottom. The water was only 100 feet (30 m) or so deep, far shallower than the average 3-mile (4.8 km) depth of the world's oceans. The mud came from the erosion of the volcanic mountain chain and the fold and thrust belt west of Yellowstone. Some of the mud formed due to the chemical breakdown of feldspar and other minerals in the rocks making up the mountains. Additional mud was formed as rivers transporting sediment to the seaway broke it into smaller and smaller particles. Unlike the white or light-gray lime mud of the Madison limestone, which was deposited in clear blue, tropical marine water (see vignette 3), the mud that washed into the Cretaceous Interior Seaway was mostly brown or dark gray due to the presence of abundant clay minerals and bits and pieces of organic debris. The water of the seaway would have been particularly muddy on its western side, close to the sources of sediment.

In contrast to the relatively deepwater conditions in which the shale was deposited, the more massive brown sandstone of the Eagle Forma-tion was deposited in much shallower water near the shoreline. Sand-rich deposits, including deltas and beaches, formed on the western margin of the seaway during Late Cretaceous time. As the seaway drained and the western shoreline drifted eastward into Yellowstone Country, these sand-rich deposits were laid down on top of the older offshore muds deposited when the water was deeper. The upward increase in the thickness and abundance of sand layers in the cliff is a direct reflection of this shal-lowing of the seaway; it is called a *regressive sedimentary sequence*. The seaward movement of a shoreline is called *regression*; it is the opposite of transgression (discussed in vignettes 1 and 2), in which a shoreline moves in a landward direction and water depth increases.

The thin beds of sandstone interbedded with the shale near the base of the cliff are called *turbidites*. If you have a pair of binocu-lars, this would be a good time to use them, although the layers can be seen without binoculars. The turbidites probably formed during large storms. For example, wind from a hurricane can push seawater onshore. Called a *storm surge*, this body of water overruns low-lying coastal regions. A memorable modern-day storm surge occurred in Louisiana during Hurricane Katrina; the surge crested up to 20 feet (6 m) above sea level and washed inland about 6 miles (10 km), and as far as 12 miles (20 km) along rivers. When such large storms abate, the storm surge flows back out to sea, carrying with it lots of sediment. In addition, more sediment is washed offshore by flooding rivers, because rain accompanying major storms typically causes rivers that empty at the shoreline to flood. Turbidity currents, turbulent mixtures of sediment and water, form in these sediment-laden retreating flows and can carry sand-sized sediment well over 10 miles (16 km) offshore. Because it is denser than seawater, the turbidity current typically hugs the sea bottom. Eventually, it slows down and deposits the sand and finer sediment as a turbidite.

Looking toward Eagle Nest Rock at stop 1 in the Gardner River canyon. The nearly vertical base of the canyon wall is shale of the Telegraph Creek Formation interbedded with thin sandstone turbidites.

Closer view of the gray shale of the Telegraph Creek Formation and its interbedded, yellowish brown sandstone turbidites near Eagle Nest Rock.

Turbidites usually occur—as they do at stop 1—as layers of sandstone interbedded in shale. Characteristic of turbidites, each layer has a nearly uniform thickness from one part of the outcrop to another. When a turbidity current flows across a relatively flat seafloor, it spreads out in all directions, depositing a layer of sand and silt that tapers very slightly toward its edge. Geologists have studied and mapped individual turbidites outside the Yellowstone region and found that they can reach lateral distances exceeding 70 miles (115 km). Due to their significant lateral extent, turbidites seen at an outcrop scale, such as those at stop 1, usually appear to have parallel upper and lower surfaces.

A great view of the entire regressive sequence of the seaway can be seen at stop 2. Looking northeast toward Mt. Everts, the long ridge to the east, you'll see an exposure of gray shale with a prominent yellowish brown cliff of Eagle Formation sandstone about halfway up the slope. Below it is the lighter-gray Telegraph Creek Formation, and above it is the mostly gray shale of the Everts Formation. After the western shoreline of the seaway passed eastward through the Yellowstone region, leaving behind the Eagle Formation, the region became part of an eastward-sloping coastal plain. Rivers meandered lazily across the plain toward remnants of the seaway, depositing more than 1,000 feet (300 m) of sediment. Much of the shale exposed in the Everts Formation was deposited in river floodplains on this coastal plain. The Everts Formation contains coal seams, which developed from plant debris that collected in swampy areas.

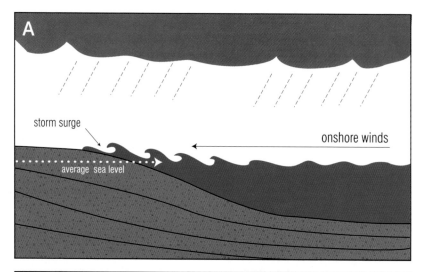

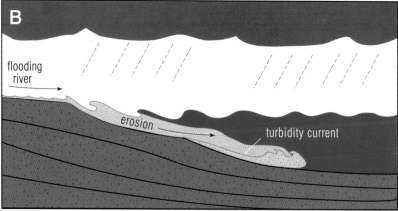

The sequence of events that may have formed the turbidites at stop 1. (A) A storm surge pushes water onshore along the western margin of the Cretaceous Interior Seaway. (B) When the storm abates, the water flows back out to sea, carrying a large amount of sand-sized sediment. Combined with sediment from flooding rivers, these flows produced turbidity currents, flowing turbulent mixtures of sediment and water that deposit a bed of sandy sediment—the turbidite.

Imagine the scene here 100 million years ago, when a major interior seaway bathed the relatively flat Yellowstone Country in murky brown water. Land-dwelling dinosaurs roamed the low-lying coastal regions. Intense tropical storms occasionally battered the shoreline, producing storm surges that swept onshore and then carried big plumes of sediment out to sea as turbidity currents, which deposited thin, sandy turbidites on an otherwise muddy bottom. Most of the time the seaway was relatively calm, and mud suspended in the water slowly settled to the bottom.

As the seaway shrank, the shoreline migrated eastward, transforming the Yellowstone region into a gentle coastal plain with swampy, vegetated floodplains separated by rivers that carried sandy sediment toward the retreating shoreline. As the Cretaceous period drew to a close, this coastal plain began to develop topographic relief, first as a series of hills, and later as mountains that reflected the onset of compressional deformation in the area. In vignettes 5 and 6, we'll take a look at some of the geologic structures and coarse-grained sedimentary deposits associated with this initial phase of mountain building that continues today in Yellowstone Country.

Looking northeast from stop 2, across the Gardner River canyon, where the regressive sedimentary sequence of the Cretaceous Interior Seaway is exposed.

5.
Mountains Reduced to Rubble
Sphinx Mountain and The Helmet

Although the Cretaceous Interior Seaway persisted in the midcontinent until about 72 million years ago, it had drifted east of Yellowstone by about 83 million years ago. For a few million years, Yellowstone Country was a broad, mostly eastward-sloping coastal plain characterized by meandering rivers and lush floodplain environments. As discussed in vignette 4, during Late Cretaceous time the western margin of the North American continent was characterized by a subduction zone, in which the oceanic Farallon Plate was plunging under the North American Plate. The compressive forces associated with the subduction caused giant slabs of North American crust to be folded and shoved eastward on top of one another along thrust faults. Called *thrust sheets*, these sections of crust overlapped each other like shingles on a roof and formed a belt of deformed rock called a *fold and thrust belt*. The belt extended from southern Canada to northern Mexico. In most of Idaho, western Montana, and western Wyoming the fold and thrust belt formed the initial topographic expression of the northern Rocky Mountains.

At first, the fold and thrust belt was located west of the Yellowstone region in Montana and Idaho. As compression continued and more thrust sheets were shoved eastward, the leading edge of the belt migrated into the western part of Yellowstone Country around 76 million years ago, profoundly changing the topography of the region west and southwest of Yellowstone National Park. Major mountain ranges and intervening valleys developed from what had been a gently sloping coastal plain. Although remnants of these mountains still exist

56

today south and west of the park, most have been deeply eroded or overprinted by more recent deformation. Nevertheless, the fact that they existed is undeniable, in part because of the presence of folds and faults that developed during the mountain building, and also because large volumes of sediment shed off the mountains were preserved. Later hardening into sedimentary rock, these sediments have provided geologists with a record of this early phase in the formation of the northern Rocky Mountains.

The central part of the Madison Range, northwest of Yellowstone National Park, contains one of the best-preserved accumulations of

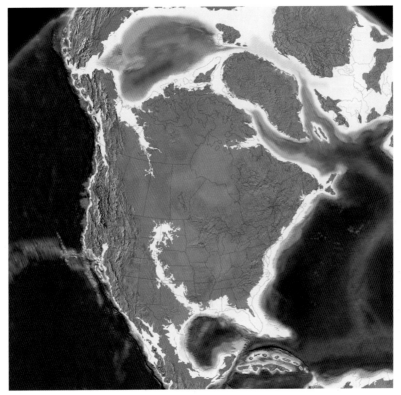

Reconstruction of North America during latest Cretaceous time, about 65 million years ago. Remnants of the Cretaceous Interior Seaway still existed in the form of two disconnected arms of marine water, one projecting north from the Gulf of Mexico and the other south from the Arctic Ocean. Mountain building in the western portion of Yellowstone Country helped force the western margin of the seaway eastward, out of Yellowstone Country. —Courtesy of Ronald Blakey and the Northern Arizona University Geology Department

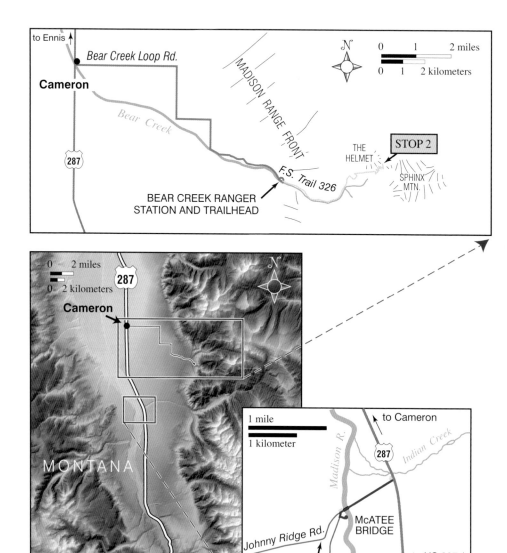

GETTING THERE

Stop 1 is a great place to view Sphinx Mountain and The Helmet from a distance. To get there from Ennis, Montana, follow US 287 south for about 18 miles (29 km) and turn right (west) onto Johnny Ridge Road. From West Yellowstone, Montana, follow

US 287 north for about 53 miles (85 km) and turn left onto the road. Proceed west on Johnny Ridge Road for about 0.7 mile (1.1 km) to the bridge over the Madison River. Cross the river and continue past the McAtee Bridge Fishing Access Site for another 0.5 mile (0.8 km) or so to the top of a steep grade, stop 1, which provides a stunning view to the northeast of the central Madison Valley and Madison Range.

Stop 2 requires a spectacular but strenuous hike to the saddle between Sphinx Mountain and The Helmet. To get there, backtrack to US 287 and turn left (north). Drive about 7.2 miles (11.6 km) to Cameron, Montana, and turn right (east) onto Bear Creek Loop Road. Follow this road approximately 7.4 miles (11.9 km); it makes a few right-angle jogs, but you want to keep heading south and east until you reach Bear Creek Road, which continues east into Bear Creek canyon. After about 0.6 mile (1 km) the road ends at the Bear Creek Ranger Station, which has a small, primitive campground. Park in the backcountry parking lot and follow the signs for Trail No. 326, the Trail Creek Trail. Be advised that Trail No. 325, the Bear Creek Trail, also has its trailhead at the Bear Creek Ranger Station, but this is not the one you want. Proceed up Trail No. 326 and take the left fork about 1.6 miles (2.6 km) from the trailhead. Continue hiking uphill on the obvious trail. You will encounter a series of long switchbacks before reaching great views of Sphinx Mountain and The Helmet about 1.3 miles (2.1 km) from the fork. From the point where you first encounter these views, it is 1 mile (1.6 km) to the saddle (stop 2).

Although the trail is well maintained and easy to follow, the hike to and from the saddle is more than 8 miles (13 km) and involves an elevation gain of 2,680 vertical feet (817 m). If you attempt to hike beyond the saddle to the summit of either Sphinx Mountain or The Helmet, be advised that the footing is loose and the conglomerate is weak and not suitable as climbing rock. These conditions, combined with the abundance of sheer cliffs with long drops, can make summit attempts very dangerous. The scramble from the saddle to either summit and back will take much of a long summer day to complete, so it is best to spend one or two nights camping in the backcountry if you plan to reach one of the summits.

sedimentary rock eroded from the early northern Rockies. Called the Sphinx Mountain conglomerate, this rock is nearly 0.7 mile (1.1 km) thick. Both the sentinel-like Sphinx Mountain and its sidekick The Helmet are composed of this conglomerate, most of which was deposited between 75 and 65 million years ago. Sculpted by glaciers over the past 2 million years into sharp mountains, these are two of the most distinctive and easily recognized peaks in the Yellowstone region. They are clearly visible from US 287 as you drive through the Madison Valley.

A conglomerate consists of marble-sized pebbles, fist- to cabbage-sized cobbles, and basketball- to microwave oven–sized (or larger) boulders stuck together with mineral cements. These natural cements can be made of various minerals, including calcite, quartz, iron oxide, and others, and typically are introduced to the sediment by groundwater after the sediment has been buried. Groundwater slowly percolates through the pore spaces in the sediment, and a variety of dissolved ions (atoms with electric charges) and molecules precipitate out of the water as minerals, binding the particles and turning loose sediment into hard rock.

Because their grain size is coarse relative to other sedimentary rocks, such as sandstone and shale, geologists generally interpret conglomerates as having been deposited relatively close to their source rock—the rock that was eroded to form the pebbles, cobbles, and boulders. This generalization is based on the observation that eroded material is broken into smaller and smaller pieces the farther it is transported from its source by rivers or other water currents, although the rate at which this process happens depends in large part on the strength and hardness of different rock types. Soft rock such as shale is broken down very quickly into sand-sized or smaller particles, whereas harder rock such as granite can be transported much farther before being broken down. Sedimentary particles that form from the physical breakdown of rock are called *clasts*, and everything from the largest boulder to the tiniest particle of clay is a clast.

Conglomerates are often deposited as a distinctive landform called an *alluvial fan*, so named because when viewed from above, it looks quite similar to an old-fashioned handheld fan. An alluvial fan has a sharp apex, its topographically highest point, and a semicircular perimeter. Typically, an alluvial fan forms where a relatively steep, fast-flowing mountain stream slows down as it crosses onto more gentle terrain, such as where a fault delineates the boundary (front) of a mountain range. The stream splits into multiple smaller streams that spread the coarse sediment across the fan's surface, from its apex to its perimeter. The Sphinx Mountain conglomerate was deposited in such a setting.

Looking northwest from stop 1, The Helmet is the mountain to the left of the larger Sphinx Mountain. Within 0.25 mile (0.4 km) of the westernmost outcrop of conglomerate making up The Helmet is a major thrust fault called the Scarface Thrust Fault. The sparsely

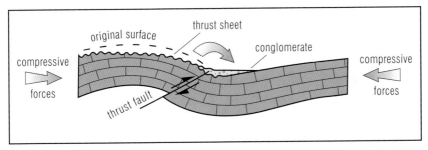

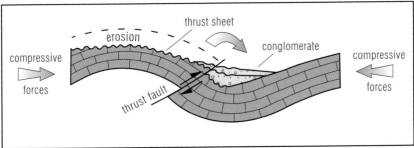

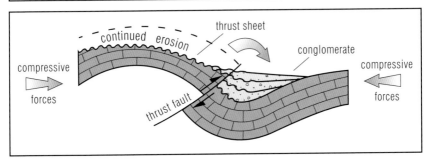

Thrust faults form when rock is fractured due to compressive stress. The compression forces large sections of crust, called *thrust sheets*, on top of one another like giant overlapping shingles along these faults. Large volumes of sediment, including an abundance of boulders and cobbles, are eroded from the elevated thrust sheet and typically accumulate in front of it. The Sphinx Mountain conglomerate was deposited in this manner.

timbered, prominent northwestern slope of The Helmet essentially is the exhumed expression of the Scarface Thrust Fault, over which the Scarface Thrust Sheet was uplifted during Late Cretaceous time, rising to an elevation above The Helmet's summit. In fact, since the streams that deposited the conglomerate of the mountains had to have run downhill, the thrust sheet must have been higher than the summit of Sphinx Mountain.

The low ridge of grayish brown Precambrian basement rock on the skyline west (left) of The Helmet is a remnant of the Scarface Thrust Sheet. The younger sedimentary strata that covered it and composed the bulk of the original thrust sheet have been eroded, with much of their erosional products forming the Sphinx Mountain conglomerate. A second major source of the conglomerate's clasts was the Shedhorn Mountain Thrust Sheet, which was uplifted along the Shedhorn Mountain Thrust Fault about 2 miles (3.2 km) southwest of the mountains. The timbered ridge immediately to the right of Sphinx Mountain is a remnant of this thrust sheet. The thrust sheets and conglomerate are

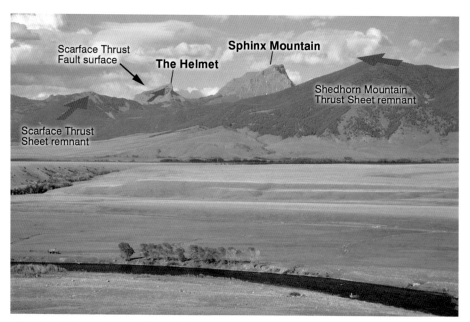

Sphinx Mountain and The Helmet as seen from stop 1. The Madison River is in the foreground. The sloping side of The Helmet facing the viewer is interpreted to be the surface expression of the Scarface Thrust Fault, over which the Scarface Thrust Sheet moved. Arrows denote the direction of movement of the thrust sheets.

no longer in direct contact because each has been substantially eroded since Cretaceous time.

A more detailed picture of the depositional history of the Sphinx Mountain conglomerate, and its relationship to the thrust sheets, can be had at stop 2—the saddle between Sphinx Mountain and The Helmet.

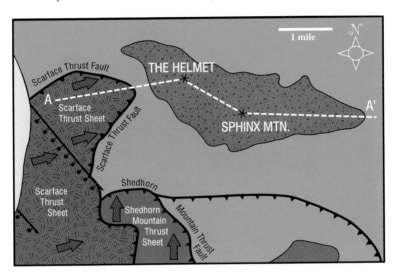

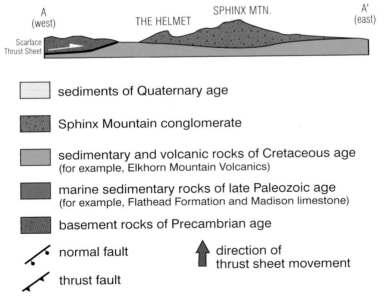

Simplified geologic map of Sphinx Mountain and The Helmet. Thrust sheets were uplifted along the Scarface and Shedhorn Mountain thrust faults, forming mountains, which eroded and produced the Sphinx Mountain conglomerate.

Although this truly is one of the finest hikes in Yellowstone Country, it is strenuous and should not be undertaken lightly.

For the first 1.5 miles (2.4 km) or so, Trail No. 326 stays mainly in the woods along Bear Creek and passes by small outcrops of Precambrian basement rock that are part of the deeply eroded Scarface Thrust Sheet. This rock is gneiss and schist, two types of metamorphic rock that form deep within the Earth when rock is heated and squeezed. If you stop to catch your breath by one of these outcrops, look for garnets—small red minerals that are common in the schist. Once past the fork in the trail, you climb through a series of beautiful meadows with impressive views of Sphinx Mountain; the best views of The Helmet come after about 3 miles (4.8 km). The last 1 mile (1.6 km) of the hike passes directly below the south face of The Helmet and provides a superb cross-sectional view of the conglomerate. Particularly well displayed is the transition from vertically tilted beds of conglomerate on the southwest corner to the nearly horizontal beds at the summit of the peak. The layers on the southwestern corner of The Helmet were bent to their vertical position by the friction created as the Scarface Thrust Sheet was shoved upward along the Scarface Thrust Fault.

As the trail passes below the south face of The Helmet, you will see that a large cliff forms the upper half of the mountain, which includes the vertically bent beds. A more gently inclined, grass-covered slope

The south face of The Helmet as seen from the trail about 0.5 mile (0.8 km) from the saddle. The grassy slope marks a transition between the two source areas of the conglomerate.

separates this cliff from a second cliff forming the lower half of the mountain. Geologists have determined that the clasts forming each cliff came from different source areas. The lower cliff contains clasts derived mostly from erosion of the Scarface Thrust Sheet, whereas the upper cliff contains a substantial contribution of clasts from the Shedhorn Mountain Thrust Sheet. How did geologists figure this out?

When rivers transport larger clasts, the flow of water at the bottom of the riverbed tends to arrange them in a stable position. Usually, flat clasts lean upstream and stack up against one another like a series of dominoes, forming a sedimentary structure called *imbrication*. By noting the orientation of the stacked clasts, geologists are able to determine the direction the rivers were flowing even though they are long gone. Imbrication in the lower cliff of conglomerate indicates that the clasts were deposited by rivers flowing from the west, from the direction of the Scarface Thrust Sheet. In contrast, imbrication in the upper cliff indicates that the clasts were deposited by rivers flowing from the south, from the direction of the Shedhorn Mountain Thrust Sheet.

Other rock types, including shale and sandstone, were deposited in the alluvial fans that formed in front of the advancing thrust sheets.

The upturned conglomerate layers *(upper portion of photo)* on the southwestern edge of The Helmet contrast greatly with the horizontal layers *(bottom of photo)*. The vertical layers were deformed by the friction of the overriding Scarface Thrust Sheet, which moved from left to right.

Cropping out next to the trail at stop 2 are sandstone beds, each about 1 foot (30 cm) thick, along with reddish brown siltstone. These relatively fine-grained rocks are part of the base of the upper cliff and were deposited close to the outer perimeter of the alluvial fans derived from the Scarface Thrust Sheet, beyond the zone where coarser clasts were transported. Above the sandstone and siltstone are thicker, more prominent cobble- and boulder-bearing beds of conglomerate. These were deposited as the alluvial fans expanded in size, or moved forward from the advancing Scarface Thrust Sheet. When these beds were deposited, the steeper apex of the fans was closer to stop 2.

Before heading back down the mountain, follow the footpath from stop 2 northward across the saddle to the base of the lowest cliff on Sphinx Mountain. The many large blocks of conglomerate lying around provide an opportunity to look at the different rock types that compose the conglomerate. When they were being uplifted, both thrust sheets consisted of intact layers of strata ranging from the Cambrian Flathead Formation to the Late Cretaceous Elkhorn Mountain Volcanics. The conglomerate includes clasts from about a dozen sedimentary

Looking west at The Helmet from the southwest flank of Sphinx Mountain, about 0.25 mile (0.4 km) east of stop 2. The saddle is just visible on the right. The prominent flat slope on the back side of The Helmet is interpreted to be the surface expression of the Scarface Thrust Fault. The timbered ridge beyond and left of The Helmet is Precambrian basement rock—a remnant of the ancient Scarface Thrust Sheet. Large blocks of Sphinx Mountain conglomerate are visible in the foreground.

rock formations, including the Cambrian-age Meagher and Pilgrim limestones, Devonian-age dolomite of the Jefferson Formation, Mississippian-age Madison limestone, and fine-grained shale and sandstone deposited in the Cretaceous Interior Seaway (see vignette 4). Geologists have correlated the different rock types of the clasts to different geological formations in the two thrust sheets.

Because both the Scarface and Shedhorn Mountain thrust sheets were eroded from the top down, as would be expected, there exists a sort of mirror image between their original layering and the sequence of clasts that were deposited in the conglomerate. It's called *stratigraphic inversion*. Clasts derived from the youngest geological formations—formerly the top of the thrust sheets—occur in the oldest, or lowermost,

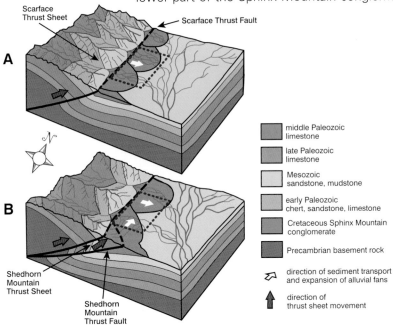

A) The Scarface Thrust Sheet was uplifted along the Scarface Thrust Fault. Clasts derived from erosion of this thrust sheet were transported eastward by rivers and deposited in alluvial fans, which became the lower part of the Sphinx Mountain conglomerate.

Scarface Thrust Sheet

Scarface Thrust Fault

A

middle Paleozoic limestone

late Paleozoic limestone

Mesozoic sandstone, mudstone

early Paleozoic chert, sandstone, limestone

Cretaceous Sphinx Mountain conglomerate

Precambrian basement rock

direction of sediment transport and expansion of alluvial fans

direction of thrust sheet movement

B

Shedhorn Mountain Thrust Sheet

Shedhorn Mountain Thrust Fault

B) The Shedhorn Mountain Thrust Sheet was uplifted along the Shedhorn Mountain Thrust Fault. Clasts derived from erosion of this thrust sheet were transported northward and deposited in alluvial fans, becoming the upper part of the conglomerate. The red box shows the approximate future position of Sphinx Mountain and The Helmet.

layers of the conglomerate; this includes clasts of the Elkhorn Mountain Volcanics. In contrast, the upper levels of the conglomerate are dominated by the much older Cambrian-age sedimentary formations that composed the lower parts of the thrust sheets. These formations were exposed once the younger rock layers of the thrust sheets had eroded away. The summit regions of both The Helmet and Sphinx Mountain are dominated by these clasts of older rock, including Flathead

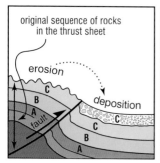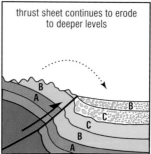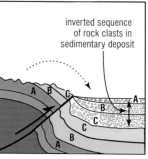

The Sphinx Mountain conglomerate is made of clasts derived from top-down erosion of the Scarface and Shedhorn Mountain thrust sheets. As a result, the lower parts of the conglomerate are dominated by clasts of the youngest geological formations in the thrust sheets, and the upper parts by clasts of the oldest sedimentary layers.

A close-up of pebble imbrication in the Sphinx Mountain conglomerate. The long axes of the pebbles mostly lean to the left, indicating that the river that deposited them flowed from left to right. Lens cap for scale.

Formation sandstone, Meagher and Pilgrim limestones, and Madison limestone.

Is it possible that the actively forming mountains now undergoing uplift and erosion in the Yellowstone area will form deposits similar to the Sphinx Mountain conglomerate? Large clasts certainly are abundant in the steep rivers draining the region, as anyone who has fished the Yellowstone region can attest, and coarse-grained alluvial fans are common, particularly along the fronts of mountain ranges such as the Madison Range. However, unlike the Sphinx Mountain conglomerate, most of this sediment is being carried away rather than forming substantial accumulations. Changing tectonics in the Yellowstone region over the past 65 million years have brought about this transition. In vignette 6, we examine the first major tectonic change that came after the Sphinx Mountain conglomerate was deposited.

A solitary horn coral preserved in a clast of Mississippian-age Madison limestone. Such fossils helped geologists determine which geological formation was eroded to form the clast. Quarter for scale.

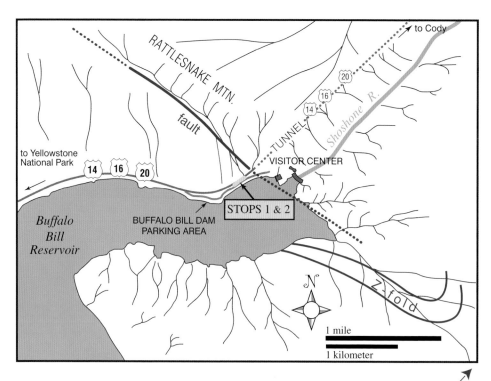

6.
Basement Rock on the Rise
Uplift of Rattlesnake and Cedar Mountains

Mountain building in Yellowstone Country first began west of the park in southwestern Montana about 76 million years ago. As discussed in vignette 5, at that time the crust was being compressed in an east-west direction as the Farallon Plate was subducted under the North American Plate. The style of deformation that formed these initial mountains was called *thin-skinned*, because most of the deformation was confined to the relatively shallow layers of sedimentary rock overlying the

---

GETTING THERE

The geologic structures discussed in this vignette can be easily seen from the parking lot of the Buffalo Bill Dam Visitor Center at the east end of Buffalo Bill Reservoir. To get there, drive 4.2 miles (6.8 km) west from the rodeo arena at the western edge of Cody on US 14/16/20 and turn left (south) into the parking lot. If you are coming from the Shoshone River canyon localities described in vignettes 1 and 2, simply drive west another 2.3 miles (3.7 km) on the highway from the intersection of Hayden Arch Road and US 14/16/20. You will pass through a tunnel before arriving at the parking lot. If you are coming from Yellowstone National Park, the visitor center is 45 miles (72 km) east of the park's East Entrance. Park near the displays of hardware associated with the dam and walk northeast along the sidewalk toward the visitor center. Stop 1—the best view of the Z-fold—is at the northeast end of the sidewalk near the bus turnaround. To reach stop 2—a good view of the Rattlesnake Mountain Fault—backtrack 50 yards (45 m) or so to the southwest.

---

crystalline basement rock. The sedimentary layers were pushed forward and upward along thrust faults that descended into the subsurface at relatively shallow angles, usually less than 15 degrees. This deformation created a north-south-trending fold and thrust belt consisting of sheets of rock shingled on top of one another.

By early Paleocene time, however, around 60 million years ago, the style of deformation changed and spread eastward through and beyond Yellowstone National Park. During this phase of mountain building, thick blocks of rock rose along reverse faults that plunged into the subsurface at much steeper angles, usually around 70 degrees. Like thrust faults, which plunge into the subsurface at shallow angles, reverse faults are the result of compressional stress. Rocks above both fault types are displaced upward relative to rocks below the faults. The steeply dipping reverse faults typically cut through both the veneer of sedimentary rock and the crystalline basement rock below. Called *thick-skinned*, this deformation style created many of the mountain ranges that still exist in the northern Rocky Mountains, including the Wind River, Bighorn, Laramie, and Medicine Bow ranges in Wyoming; the Beartooth Mountains of Montana and Wyoming; the Uinta Mountains in Utah; and the Black Hills of South Dakota.

What caused the leading edge of the deformation to sweep eastward about 400 miles (640 km)? And why did the dominant style of deformation change from thin-skinned to thick-skinned? These questions have fascinated geologists for decades. Many have suggested that the changes occurred because of a decrease in the steepness of the angle at which the Farallon Plate was being subducted, although the reasons for shallowing of the subduction of such a rock slab are not well understood.

One possibility is that the temperature of the Farallon Plate increased. As the spreading center from which the oceanic Farallon Plate was forming drifted eastward toward the subduction zone, younger and younger crust was being subducted beneath the North American Plate. Newly formed oceanic plate is hotter and less dense than old, cold oceanic plate, and thus it is more buoyant. The portion of the Farallon Plate closest to the spreading center may have been buoyant enough to rotate upward, pressing against the base of the overriding North American Plate like a giant beach ball trapped beneath the hull of a flat-bottomed boat. This upward pressure could have forced thick chunks of the North American lithosphere to rise upward along high-angle reverse faults.

Another recently published interpretation posits that an oceanic plateau was subducted as part of the Farallon Plate, effectively increasing the thickness of the plate. Where the plate was thicker, its lower

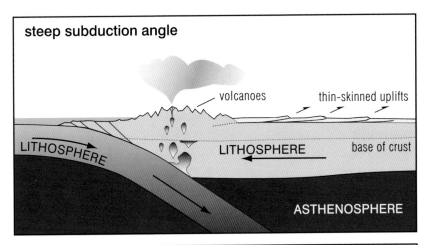

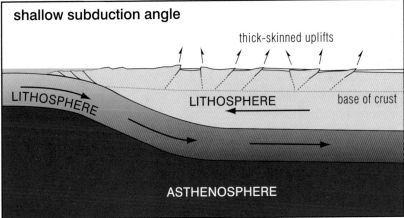

Early mountain building in the Yellowstone region was the result of two different styles of compressional deformation. The thin-skinned style involved shallow faults that were largely confined to the layered sedimentary rocks above the crystalline basement. This deformation occurred when the Farallon Plate was subducted beneath the North American Plate at a steep angle. The thick-skinned style involved reverse faults that cut more steeply downward and into the crystalline basement rock. This style likely occurred when the Farallon Plate was subducted at a shallower angle. Arrows denote direction of movement.

parts were subjected to more pressure than usual, and basalt there was converted to a more dense rock called *eclogite*. The increased density caused the Farallon Plate to sink more steeply and the overlying, attached North American Plate flexed downward with it. Eventually, the Farallon Plate and North American Plate separated, and the North American Plate rebounded upward like a beach ball that had been held underwater. As it rebounded, the plate broke into different segments along reverse faults.

Regardless of what caused it, thick-skinned deformation began around 60 million years ago in Yellowstone Country. One of the uplifts it produced is the Beartooth Uplift, a huge block of crust that stretches from Cody, Wyoming, north and west to Red Lodge, Montana. The uplift consists of crystalline metamorphic and igneous Precambrian basement rock—2.8-billion-year-old granite and gneiss that originally formed deep below the surface. This rock was shoved upward along the Beartooth Fault, a major reverse fault that is exposed in a few places along the eastern and northern fronts of the Beartooth Mountains. The Beartooth Uplift itself is broken up by several smaller reverse faults, creating a series of smaller blocks of crust. One of these blocks comprises Rattlesnake and Cedar mountains, which loom over Buffalo Bill Reservoir near Cody, Wyoming.

Construction of Buffalo Bill Dam in the upper end of the Shoshone River canyon began in 1905. It took five years to complete, and at the time it was the tallest dam in the world, reaching 325 feet (100 m) above the canyon floor. The dam was made 25 feet (7.5 m) taller in 1993 to increase Buffalo Bill Reservoir's storage capacity. Engineers were able to use strong, crystalline Precambrian rock of the canyon as the dam's foundation. Although the engineers certainly were aware of the strength of the basement rock and therefore its suitability as a foundation, it's not clear if they were aware of the fault beneath the dam. It was along this fault, called the Rattlesnake Mountain Fault, that the seemingly solid basement rock was uplifted to its present elevation. Based on analysis of the fault geometry and the matching of rock layers that have been displaced by the fault, geologists estimate that the rocks forming Rattlesnake Mountain rose as much as 8,000 feet—about 1.5 miles (2.4 km)—relative to the same layers of rock underlying the broad valley to the west that contains the reservoir.

Looking southeastward across the reservoir from stop 1, you'll see that the dam is built upon pinkish gneiss and granite, the crystalline Precambrian basement rock examined in vignette 1. To the right of the dam, the pinkish rock changes to layered sedimentary rocks that are timbered and stretch all the way to the southeastern skyline. These sedimentary layers include the Flathead, Gros Ventre, and Gallatin formations discussed in vignettes 1 and 2, as well as younger Paleozoic limestones. The sedimentary layers are bent into a giant Z-shaped fold. This fold developed as the basement rock was uplifted along the Rattlesnake Mountain Fault, which is not visible on the far side of the reservoir because it does not reach the surface. The fault runs parallel to the northwest-southeast trend of Rattlesnake and Cedar mountains.

Although the Rattlesnake Mountain Fault does not reach the surface, at least one splay of the reverse fault can be seen at stop 2. Look north across the eastbound lane of US 14/16/20 and you'll see that

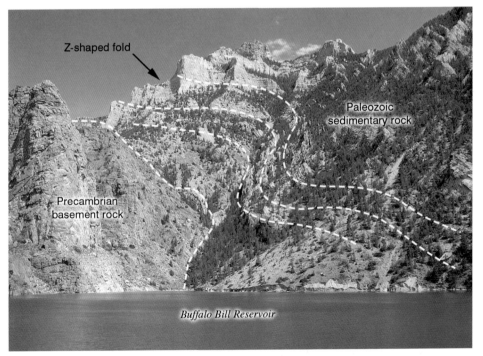

Looking southeast across Buffalo Bill Reservoir from stop 1. The pinkish brown Precambrian basement rock *(left)* and overlying light-gray Paleozoic sedimentary rock on the skyline above it *(right)* were folded and uplifted due to movement on the Rattlesnake Mountain Fault. The Z-shaped fold *(highlighted)* formed as the uplifted block of basement rock caused the overlying sedimentary layers to bend.

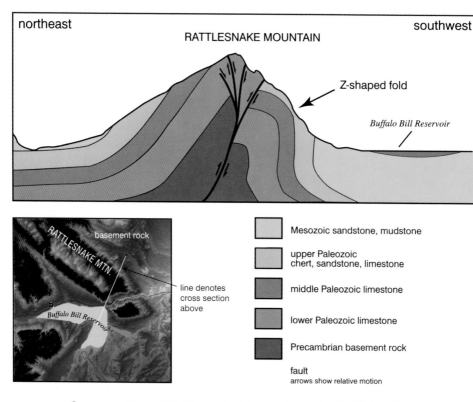

Cross section of Rattlesnake Mountain at the Buffalo Bill Reservoir, looking to the southeast from stop 1.

the crystalline Precambrian basement rock was displaced up and over the grayish green, well-layered sedimentary rocks of the Gros Ventre Formation. These rock layers are tilted up on end—they are nearly vertical. The basement rock forced them into this position as it rose along the fault splay. The Gros Ventre Formation and basement rock are normally not in direct contact with each other, as they are here. In places where the crust has not been cut by faults, the Flathead Formation separates the two.

Today, the crust in this region is no longer being compressed, nor is Rattlesnake Mountain rising relative to the valley containing Buffalo Bill Reservoir. Instead, this region is slowly being pulled apart by extensional stress—the tectonic opposite of compression. If this extensional stress continues, there is some risk that there will be a new phase of motion along the Rattlesnake Mountain Fault. This movement could weaken the dam or, in the worst-case scenario, cause it to fail.

The impressive Z-shaped fold and evidence of uplift of the rocks making up Rattlesnake and Cedar mountains provide some sense of how powerful the tectonic forces were that formed the thick-skinned uplifts in this part of Yellowstone Country. These forces are especially impressive when you consider that the Precambrian basement rock forming the core of these mountains is only a small segment of the massive Beartooth Uplift, which extends north and west for more than 60 miles (97 km).

About 5 million years after the thick-skinned deformation ceased, the tectonic setting of the Yellowstone region changed dramatically, to one marked by widespread volcanic activity. In vignettes 7, 8, and 9, we examine this phase of Yellowstone's geologic history.

Looking north from stop 2. The high wall of pink Precambrian basement rock was thrust upward against limestone and shale layers of the Gros Ventre Formation, pushing them into a vertical position in the process. The reverse fault—the contact between the two rock types—is one of several splays that merge below the surface to form the Rattlesnake Mountain Fault.

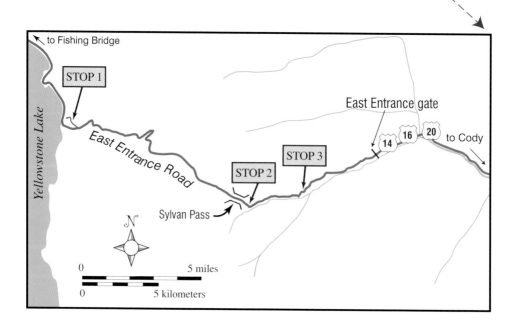

7.
Strange Brew
The Unusual Volcanic Rocks near Sylvan Pass

Most folks would agree that Yellowstone Country is famous for its history of massive, explosive volcanism. Much of this well-deserved fame comes from the three violent and enormous eruptions of the Yellowstone Volcano that occurred during the past 2 million years. But Yellowstone was home to volcanoes long before that. Between about 55 and 48 million years ago, during the early part of Eocene time, large volcanoes dominated the landscape and spewed out huge volumes of lava, rock, and ash. Between eruptions, some of this erupted volcanic

GETTING THERE

The stops in this vignette are located along East Entrance Road in Yellowstone National Park. Stop 1 is located on the road to the Lake Butte overlook. The turnoff to the overlook is 17 miles (27.4 km) west of East Entrance or 9.5 miles (15.3 km) east of Fishing Bridge on the north side of the road. Drive uphill 1 mile (1.6 km) to the end of the road, where the overlook is located, and turn around and backtrack to a very small pullout, stop 1, on the west (right) side of the road and opposite a prominent roadcut of light-gray rock. To reach stop 2, proceed east on East Entrance Road for 10.3 miles (16.6 km), up and over Sylvan Pass, to a small pullout at the first bend east of Sylvan Pass, about 0.75 mile (1.2 km) beyond the pass and on the south side of the road. To reach stop 3, continue another 2.5 miles (4 km) east of stop 2 to the large pullout on the left (north) side of the road. After parking, proceed on foot to the basalt roadcut at the west end of the pullout.

material mixed with water from storms—and possibly the hydrothermal systems associated with the volcanoes themselves—to produce powerful slurrylike debris flows that rushed down the sides of the steep mountainsides, leveling entire forests in the process. The layers of sediment deposited by these debris flows (see vignette 8) later hardened as sedimentary rock that commonly contains very well-preserved petrified wood (see vignette 9).

Collectively, the alternating layers of volcanic and sedimentary rock left behind by this long-lived series of volcanoes, along with igneous rocks formed from cooling magma beneath the volcanoes, make up the Absaroka Volcanic Supergroup. This supergroup dominates the modern-day surface geology of the northern and eastern parts of Yellowstone National Park and adjacent areas, covering about 30 percent of the park. In addition, these rocks extend southeast of the park for more than 50 miles (80 km) into northwestern Wyoming. The supergroup is estimated to contain about 7,200 cubic miles (30,000 cubic km) of rock—enough to form a layer about 2 inches (5 cm) thick over the surface of the entire Earth. The supergroup consists of three different groups of rock. From oldest to youngest, these are the Washburn Group, Sunlight Group, and Thorofare Creek Group. Each group contains rock of volcanic and sedimentary origin, as well as igneous rock that solidified underground.

The crust and upper mantle below Yellowstone partially melted to form the magma that became the supergroup. Once generated, the magma rose slowly toward the surface, partly melting the overlying rock and partly forcing its way upward through fractures in it. Although much of the magma reached the surface and erupted from the volcanoes, forming volcanic rock, some of the magma cooled and solidified underground, though probably not more than 1 or 2 miles (1.6 to 3.2 km) below the ground surface during Eocene time. The rock that cooled underground is called *intrusive igneous rock* or *plutonic igneous rock*, the latter a reference to Pluto, the Roman god of the underworld. In Yellowstone these bodies of intrusive igneous rock, generally called *plutons*, represent the now-exposed internal parts of the Eocene volcanoes.

The Absaroka Volcanic Supergroup appears to have come from four main sets, or clusters, of volcanoes. One cluster was located in and around the remote northwestern corner of Yellowstone National Park in the central and southern Gallatin Range. Another set was located in the north-central part of the park and included Mt. Washburn, which is one of the old volcanoes. The third set was located mostly east of the park and centered at Sunlight Peak, the remnants of Sunlight Volcano. The fourth set was situated mostly southeast of the park along the crest of the rugged Absaroka Range, which extends well out of the park into northwestern Wyoming. The volcanoes within each cluster had classic cone shapes that built up over multiple eruptions as magma, ash, and volcanic rock were extruded and erupted from a centralized vent. The layers of volcanic rock that resulted were interstratified with layers of sedimentary rock that formed as rock, sand, and mud were moved down the steep slopes of the volcanoes and deposited.

At the scale of the entire Yellowstone region, the volcanoes were hemmed in on the north and east by the Beartooth Uplift and on the west by mountains that had risen due to thrust faulting nearly 15 million years earlier (see vignette 5). At a more local scale, dense forests blanketed the steep sides of the volcanoes and the valley floors between them, though the summits of the largest may have been above tree line. The biggest volcanoes were tens of miles across at the base and probably reached elevations well over 10,000 feet (3,050 m) above sea level. They must have been majestic features, beautiful to behold.

The original volcanoes have been deeply eroded and in general are no longer easily recognizable. However, we know where many of them were located because the intrusive rock that cooled below them is now exposed, in some cases forming prominent mountains in the Yellowstone region. Examples include Sunlight Peak, located in the remote backcountry about 10 miles (16 km) east of Yellowstone's eastern boundary; Meldrum Mountain and Black Butte in the Gallatin Range; Mt. Washburn in the middle of the park; Lake Butte just east of Yellowstone Lake; and Colter Peak in the southeastern part of the park.

Considerable debate exists among geologists concerning the origins of the Absaroka Volcanic Supergroup, and many questions remain. Much of the discussion is focused on what the tectonic setting must have been like during Eocene time to create the supergroup's volcanic rocks. Some geologists have suggested that the magma that formed the supergroup was derived from partial melting of the Farallon Plate, which was being subducted beneath the North American Plate. Others have speculated that the Farallon Plate followed an older plate into the subduction zone, and that a break, or window, between the two spreading plates caused mantle from the asthenosphere to well up against the base of the North American lithosphere, resulting in a strange magmatic brew with an unusual composition. Regardless of the uncertainty regarding the tectonic setting during Eocene time and its relationship to the Absaroka Volcanic Supergroup, geologists do agree that the supergroup—now deeply eroded by glaciers—provides a chance to examine the internal plumbing system of the ancient volcanoes.

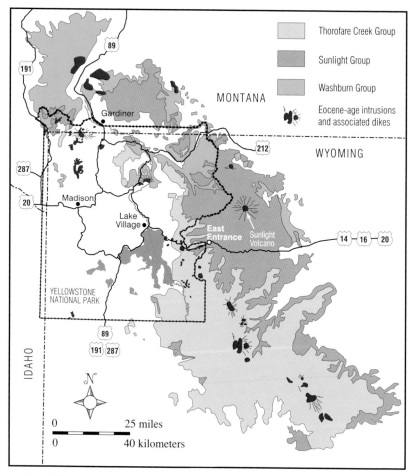

Simplified geologic map of the Absaroka Volcanic Supergroup, which is subdivided into—from oldest to youngest—the Washburn, Sunlight, and Thorofare Creek groups and includes intrusive igneous rocks that formed below the surface from cooling magma.

We'll start our examination of Eocene igneous rocks at stop 1. The light-gray rock at this stop is andesite, an igneous rock that contains abundant feldspar. It formed between 48 and 45 million years ago in the underground plumbing system of a volcano associated with the Thorofare Creek Group, the youngest rocks of the Absaroka Volcanic Supergroup. Between Lake Butte and Sylvan Pass there are several small plutons made of rock similar to this andesite. The biggest of the plutons have a rough diameter of about 1 mile (1.6 km); each probably supplied magma to a different volcanic vent.

The andesite exposed on the east side of the road is massive—that is, not layered like sedimentary rock. The massive character reflects the fact that the magma that formed the rock cooled slowly and congealed as one solid mass. You'll also notice that the andesite contains a series

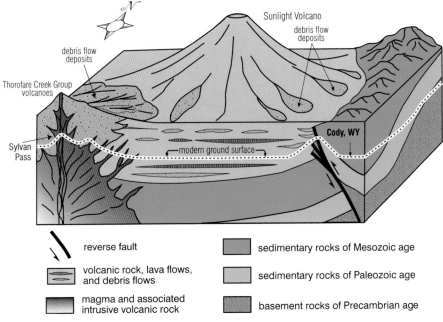

Reconstruction of the region around Sunlight Volcano about 45 million years ago, showing the approximate position of the modern-day ground surface along US 14/16/20 between Sylvan Pass and Cody, Wyoming. The cone-shaped Sunlight Volcano, about 50 miles (80 km) wide at its base, dominated the landscape and produced most of the rock of the Sunlight Group. Later, a cluster of smaller volcanoes associated with the Thorofare Creek Group formed on the southwestern flank of Sunlight Volcano. Their vents were located between Lake Butte and Sylvan Pass.

of prominent cracks, called *joints*, which are roughly parallel to each other. The most prominent joints dip steeply and slope from the upper right of the outcrop to the lower left. The flat, steeply sloping surfaces of the outcrop represent one side of a joint; the rock formerly on the other side of the joint was removed through erosion or as the road was constructed. When massive, unlayered rock such as this undergoes tectonic deformation, it typically cracks in a nonrandom fashion, forming one or more sets of parallel joints. In this case, the joints resulted from tectonic activity that has taken place in the Yellowstone area since Eocene time.

If you have a closer look at a piece of the andesite, you'll see that the rock consists of three basic components. Most obvious are the black crystals, most of which are about ⅛ inch (3 mm) across. Just about all of these have one or more flat sides, and many are rectangular. Some have a hexagonal shape, although typically the hexagon is a bit squashed, so not all of the sides have the same length. Many of the minerals are marked by a flat, shiny surface. These black crystals are the minerals amphibole and biotite, both iron-rich silicate minerals common in igneous rocks. In addition, there is an abundance of white crystals, most of

The andesite at stop 1 cooled and solidified underground within a volcano associated with the Thorofare Creek Group. The prominent cracks and flat surfaces are joints that formed due to more recent tectonic compression.

which are square or rectangular. These are the mineral plagioclase, a type of feldspar. The relatively large size of the amphibole, biotite, and plagioclase crystals reflects the fact that these minerals had ample time to grow in the magma as it slowly worked its way upward through the crust. In contrast, the light-gray matrix in between these large crystals is finely crystalline. It reflects the relatively rapid final stages of cooling that occurred in the pluton after the magma had intruded relatively shallow levels of the crust.

If you look closely, you will see that some of the plagioclase crystals are grayer in the center and whiter at the edges. Although it is subtle and most easily seen using a magnifying lens, this color variation reflects the fact that the composition of the plagioclase changed as the crystals grew in the magma. The early-formed centers are calcium rich, whereas the later-formed edges are sodium rich. This change, called *compositional zoning*, occurred because the crystals that formed first incorporated calcium into their structure but left sodium in the magma. This depleted the magma of calcium and enriched it in sodium, so as the plagioclase crystals continued to grow, they incorporated more sodium and less calcium into their structure.

Close-up of andesite at stop 1. Both the black amphibole and biotite and white plagioclase crystals had been growing in the magma for a while before it cooled and solidified. In contrast, the finely crystalline, light-gray matrix between represents the last bit of magma, which cooled relatively quickly. Dime for scale.

The rocks at stop 1 cooled within magma chambers that ultimately fed volcanoes associated with the Thorofare Creek Group. In addition to these plutonic igneous rocks, there are dozens of igneous dikes that radiate away from the volcanoes. They also reflect the subterranean cooling of magma that never erupted at the surface. In contrast, the rock at stop 2 formed on the surface close to the vent of one of the volcanoes.

From the small pullout at stop 2, proceed cautiously across the road to the roadcut on its north side. Called *volcanic breccia*, the rock exposed in the roadcut is made of angular fragments (clasts) of volcanic rock set in a finer-grained matrix. Most of the angular fragments are made of the same rock type: a finely crystalline, yellowish brown rock with small black crystals of amphibole. In addition to these, there are angular fragments of light-gray volcanic rock also containing amphibole crystals, as well as a few fragments of dark volcanic rock. In some parts of the outcrop it is difficult to distinguish the light-gray matrix from clasts of the same color because they are the same type of rock.

Although volcanic breccias are very common within the Absaroka Volcanic Supergroup, most are characterized by crude layering and a greater diversity of clast types. In contrast, the rock at stop 2 does not appear to be layered, and nearly all of the clasts are made of the

The strikingly angular shape of the clasts and the fact that most are composed of the same volcanic rock type and encased in a finely crystalline volcanic matrix suggests that the breccia at stop 2 came from an explosive volcanic eruption. Lens cap for scale.

same yellowish brown volcanic rock, suggesting they had a common origin. One interpretation is that the clasts are bits and pieces of the same source rock that existed in or near a volcanic vent. The very angular shape of the fragments indicates that they were not transported to this site by streams, which would have rounded them, but rather were produced by fragmentation during an explosive eruption. The few darker clasts, composed almost completely of amphibole, probably formed deep below the surface and were transported upward through

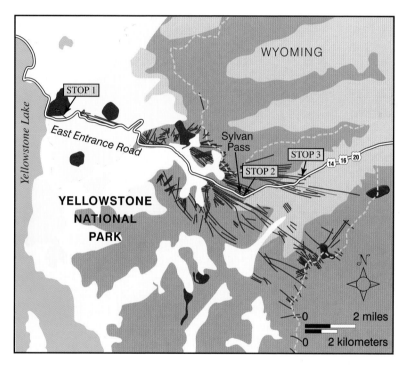

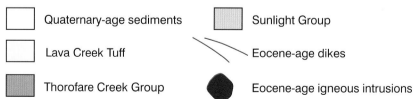

Simplified geologic map of the region near Sylvan Pass. Several small, roughly circular igneous intrusions represent magma chambers that formed below volcanoes associated with the Thorofare Creek Group. Each of the large intrusions probably formed under a different volcano. Some of the magma was injected into a series of radial fractures that projected outward from the volcanoes and cooled to form the dikes.

the volcano during an eruption. The light-gray matrix, or *groundmass*, between the clasts likely formed from volcanic ash that erupted at the same time that the clasts were fragmented. As the ash and clasts moved downslope away from the vent, they mixed together, forming this rock.

As indicated by the rocks exposed at stop 1 and the dozens of igneous dikes that radiate away from the Sylvan Pass area, it's clear that some of the magma that formed portions of the Absaroka Volcanic Supergroup never erupted at the surface. Much of the magma, however, actually did erupt, producing lava flows. A superb example of a lava flow associated with Sunlight Volcano is visible at stop 3. From the roadside pullout, proceed carefully on foot to the prominent roadcut immediately to the west, around the corner and upslope.

The rock exposed in the roadcut is an iron-rich rock called *basalt*. It is part of the Sunlight Group, which contains rocks that are a bit older than the Thorofare Creek Group. This basalt is striking because it is chock-full of small, rounded, white, pink, and red features that are typically 0.25 to 0.5 inch (6 to 12 mm) long and commonly elongate. Called *amygdules*, they are bubble holes that formed in the basalt and were later filled in by a variety of minerals. When the basalt erupted onto the surface, gas that had been dissolved in the lava because it was under

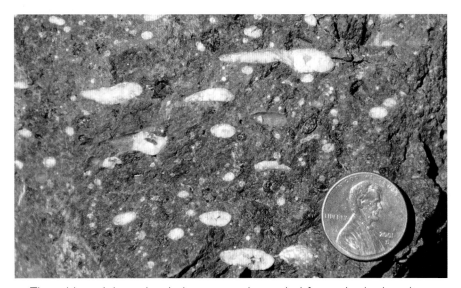

The white, pink, and red elongate and rounded forms in the basalt at stop 3 are amygdules, bubble holes that formed in the basalt and were later filled with minerals. The rough alignment and elongate shape of many of them reflect that the bubble holes were stretched before the basalt flow solidified. The dark square objects are crystals of pyroxene. Penny for scale.

pressure came out of solution. A similar process occurs when a can of beer is opened (pressure is released) and carbon dioxide gas that had been dissolved in the beer comes out of solution, forming tiny bubbles.

The fact that many of the amygdules have an elongate form and are parallel to other amygdules reflects stretching of the bubbles before the lava solidified. This stretching occurred as the flow moved its last few inches down the side of Sunlight Volcano. Most of the minerals filling the former bubble holes are zeolites, a group of minerals characterized by unusual cagelike molecular structures. The dimensions of the molecular cage of some zeolites is very specific and allows the minerals to serve useful functions in the pharmaceutical and chemical industries, for example, being used as filters for other molecules with specific sizes and shapes.

The basalt represents the partial melting of rock that existed far below the surface, in the deep lithosphere or possibly the upper asthenosphere. Without detailed chemical analysis, it is not possible to say with certainty from how far down the magma that formed this basalt was derived, or how the magma evolved as it made its way toward the surface, but basalt erupted on continents is commonly derived from partial melting of the upper mantle. Look for dark green crystals that are about 0.25 inch (6 mm) across and have a square, hexagonal, or stubby rectangular shape. These are pyroxene, a mineral commonly found in basalt.

Although the andesite, volcanic breccia, and basalt at this vignette's stops are only a few of the rock types that make up the Absaroka Volcanic Supergroup, they provide an idea of the diversity of magmas that formed this immense volume of volcanic rock. In large part, this diversity of rock types and the mineral crystals they contain has provided fuel for the ongoing controversy regarding the tectonic setting that led to the formation of the supergroup. As geoscientists learn more about the supergroup, more informed interpretations regarding this important chapter of Yellowstone's geologic history no doubt will emerge. Until then, however, the strange brew that formed deep within Earth and cooled to form the rocks of the supergroup will continue to be an exciting area of research, discussion, and controversy.

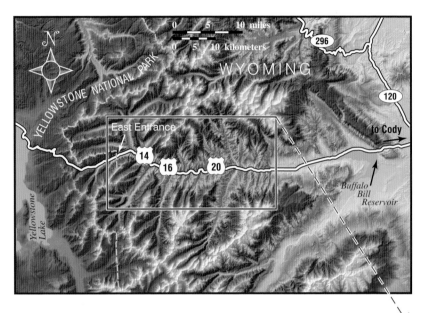

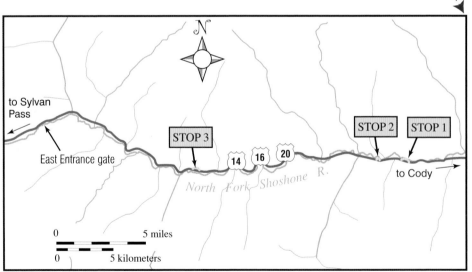

8.

Debris Flow Deposits
Coarse Conglomerate between Cody and East Entrance

Rocks belonging to the Absaroka Volcanic Supergroup are widespread across the eastern and northern parts of Yellowstone National Park, where volcanoes were located during Eocene time (see vignette 7). Included within the supergroup are several formations containing large quantities of conglomerate—a coarse-grained sedimentary rock type made mostly of fragments of older rock. These fragments, called *clasts*, range in size from pieces as small as a pea to blocks as big as a minivan or larger. Streams and other types of surface flows carried the pieces

GETTING THERE

Debris flow deposits of the Absaroka Volcanic Supergroup can be viewed between Buffalo Bill Reservoir and Yellowstone National Park's East Entrance. To reach stop 1 from the east, start at the Buffalo Bill Dam Visitor Center parking lot, which is about 7 miles (11.3 km) west of Cody, Wyoming, at the east end of the reservoir. Drive west along US 14/16/20 for 19.6 miles (31.5 km) to the roadside pullout on the right (north) side of the road. From stop 1, continue west for 1.2 miles (1.9 km) to stop 2, another roadside pullout on the north side of the road. Walk east along the road about 400 yards (365 m) and cross to the other side. You're looking for a prominent roadcut in which a large, grayish dike cuts through dark brown debris flow deposits. Stop 3 is a must-see, up-close look at several debris flow deposits. To get there, continue west for another 12.5 miles (20 km) to Chimney Rock, located on the north side of the road and marked by a sign. If you are coming from Yellowstone National Park, Chimney Rock is 11.4 miles (18.3 km) east of East Entrance.

of volcanic rock down the steep slopes of the volcanoes. The clasts accumulated around the base of the volcanoes as a deposit called an *alluvial apron*. Over geologic time, the sediments hardened.

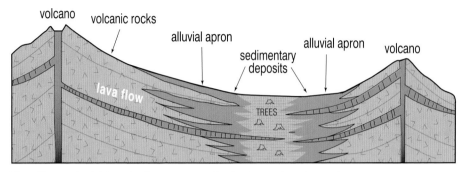

The Absaroka Volcanic Supergroup includes volcanic rock associated with the Eocene volcanoes, as well as extensive sedimentary deposits derived from erosion of the volcanoes. Most of the sedimentary deposits are coarse-grained conglomerates that covered the lower flanks of the volcanoes, forming an alluvial apron. When the volcanoes erupted, lava overran parts of the apron, producing an interlayering of volcanic and sedimentary rocks.

Among the best places to examine the sedimentary deposits of the Absaroka Volcanic Supergroup is along US 14/16/20 between Cody, Wyoming, and Yellowstone National Park's East Entrance. Along this beautiful stretch of highway, which parallels the Shoshone River, visitors can view spectacular exposures of the distinctly brown outcrops of conglomerate and sandstone that were deposited as part of an alluvial apron along the south side of Sunlight Volcano—one of Yellowstone's major volcanoes during Eocene time. The central vent of Sunlight Volcano was about 20 miles (32 km) north of US 14/16/20 in the remote backcountry of the Absaroka Range. The volcano was active between about 50 and 48 million years ago and dominated the landscape. Much of the rock forming the Sunlight Group—the middle of the three groups of rock that compose the Absaroka Volcanic Supergroup—came from Sunlight Volcano.

One of the first things visitors will notice is the impressive thickness of the sedimentary layers exposed along this 50-mile (80 km) stretch of highway. Several thousand feet of coarse sediment were deposited during Eocene time, indicating that the volcanoes had significant topographic relief. Sunlight Volcano alone was roughly 25 miles (40 km) in diameter, probably exceeded 10,000 feet (3,050 m) in elevation, and had steep flanks. Its size and general cone shape were comparable to Japan's Mt. Fuji.

Stop 1 provides an excellent cross-sectional overview of the deposits of Sunlight Volcano's alluvial apron. The Shoshone River, as well as the ancestral versions of this river and its tributaries, cut the canyon in the brown sedimentary beds. Agents of erosion have exploited vertical cracks in the rock, leaving steep cliffs where large blocks of rock simply fell away from the canyon walls. In places where vertical cracks intersected, isolated columns of rock were left standing after the rock around them had eroded away. These isolated spires, rounded from weathering, are called *hoodoos*. You can see a small army of these sentinel-like columns along the skyline to the north, directly across the river.

The steep cliff face below the hoodoos contains particularly large clasts. The largest are the size of an armchair, and many are the size of a basketball. The outcrop is also layered. The layering is defined by a

The remarkable hoodoos seen from stop 1 formed as debris flow deposits were eroded.

series of sedimentary beds that typically are around 3 to 4 feet (about 1 m) thick. Most of the individual beds are the deposit of a single debris flow—a slurrylike mixture of water and ash-rich sediment that flowed down the southern slope of Sunlight Volcano. The large size of the angular clasts throughout the outcrop indicates that the debris flows were powerful.

Debris flows belong to a class of geologic flows known as *sediment gravity flows*. Unlike flows of wind and water, which can deposit the sediment they carry in gravel bars or sand dunes respectively, a sediment gravity flow is a mixture of sediment and water that overcomes the internal forces that hold it in place on a slope, causing the whole mass to flow as a single fluid. Debris flows commonly develop on the sides of volcanoes as rain or snowmelt mixes with fine, loose volcanic ash. As the mixture flows downslope, it collects more ash-rich sediment in a process called *sediment bulking*. The resulting fluid becomes more dense and viscous—that is, thicker in terms of consistency. Many of the Eocene debris flows probably had a consistency equal to or thicker than that of wet cement, providing the flows with tremendous power—enough to pick up and move very large rocks, woody debris, and entire trees.

Because debris flows don't sort the sediment they are carrying very effectively, in contrast to flows such as rivers, deposits built by multiple debris flows tend to be crudely layered. In many cases, the top of one debris flow deposit may be difficult to distinguish from the base of the overlying deposit, although this is not always the case. For example, soil might form on top of a debris flow deposit before the next flow occurs, creating a visible bedding surface between the two flows. Layering is also visible when two flows contain sediment of substantially different grain size. For example, a debris flow deposit of only pebble-sized clasts would be easy to distinguish from one containing boulders.

Another factor contributing to the crude layering seen in debris flow deposits is the fact that debris flows usually don't cut deep channels. Relative to the much more consistent flow of a river, which, over time, is capable of eroding a deep, distinct channel into underlying rock or sediment, debris flows occur episodically. Many years can elapse between debris flows, and each event may last only a few minutes or hours. Their episodic, short-lived nature leaves them little time to erode much of a channel into underlying rock or sediment. Rather, they tend to spread out over the land surface, leaving behind a lobe-shaped deposit with a convex upper surface and a relatively flat lower surface.

Although many debris flows develop through sediment bulking, others occur when water-saturated ground fails en masse and slides downhill. A period of particularly wet weather might be enough to trigger such a slide, as can thermal groundwater heated by or derived from magma near the ground surface. The chemistry of thermal water alters and weakens rock located on the steep flank of a volcano and can cause it to fail. The failing slope can quickly evolve into a thick, powerful debris flow.

At stop 1 there are igneous dikes cutting through debris flow deposits in the westernmost (left) part of the outcrop. Because these dikes cut across the deposits, they are clearly younger. They formed as pressurized magma beneath Sunlight Volcano was injected into fissures and cracks in the sides of the volcano. The magma forming the dikes would have heated the groundwater in the debris flow deposits, forming a local hydrothermal system.

Nearly vertical dikes of igneous rock cut across the gently tilted debris flow deposits at stop 1.

For a closer look at a single igneous dike, proceed to stop 2. Once there, you might want to view the dike, the light-gray rock that cuts across the dark brown conglomerate, from the north side of the road first. Although the layering in the debris flow deposits is so crude that individual beds are difficult to distinguish, they slope from upper right to lower left, in contrast to the dike, which slopes from upper left to lower right. About halfway up the outcrop, the dike has been offset about 3 feet (roughly 1 m) by a small fault. This dike and others form a radial pattern around the former center of Sunlight Volcano.

Now carefully cross to the south side of the road. In contrast to the conglomerate, the dike is a finely crystalline igneous rock. Its color and association with similar dikes in the area suggest that it is andesite. In specimens you can pick up, the rock looks homogenous—it lacks the obvious larger crystals seen in the plutonic igneous rocks observed in vignette 7. The magma that formed this dike evidently did not contain large mineral crystals when it intruded the conglomerate. The relatively thin nature of the dike, and the fact that the magma intruded relatively cold sedimentary rocks, caused the magma to cool very quickly, so there wasn't time for large crystals to form once it had intruded the conglomerate.

There are stripes along the margins of the dike that are roughly parallel to it. These stripes developed as the dike rock was altered by groundwater flowing through a series of subtle cracks. In the upper, more deeply weathered portions of the outcrop, some of the cracks can be easily seen. As the intruding magma cooled and contracted, from the dike's margins inward, the margins cooled faster than the center of the dike. Since the margins contracted more quickly, cracks formed. Much of the alteration by groundwater probably occurred while the magma was cooling, as groundwater in the surrounding debris flow deposits was heated and circulated through the cracks.

Stop 3 is an exceptionally good place to see the details of several debris flow deposits because the outcrop is relatively unweathered, unlike the hoodoos seen at stop 1. Chimney Rock is the tall, flat-sided spire of rock on the north side of the road and immediately in front of the small parking area. An example of a weathered outcrop, in which individual layers are much less distinct, occurs immediately to the right of Chimney Rock. The weathered outcrop is more typical of debris flow deposits in the area.

A bed of sandstone, about 5 feet (1.5 m) thick, is exposed at eye level at Chimney Rock. Sandstone is a relatively minor rock type in those parts of the alluvial apron that were closest to the Eocene volcanoes, although it increases in abundance farther from the volcanoes. Although the sandstone lacks large clasts, it contains a lot of mud between its sand grains, suggesting deposition in a muddy fluid that geologists call a *hyperconcentrated flow*. Hyperconcentrated flows have a consistency that is between that of pure water and a slurrylike debris flow. Although their viscosity is not as high as those of debris flows, which are capable of rafting large boulders along, the viscosity of hyperconcentrated flows is high enough that their deposits don't usually contain crossbeds or ripple marks, which are more characteristic of pure water flowing over a bed of sand. Rather, hyperconcentrated flow deposits tend to be massive or characterized by faint, flat internal bedding, as seen here.

Two distinct beds of conglomerate rest directly on top of the sandstone. Each bed is about 6 feet (1.8 m) thick and is an individual debris

An igneous dike cutting through the sedimentary rocks of the Absaroka Volcanic Supergroup at stop 2. The subtle striped pattern seen in the lower part of the dike developed as the rock was chemically altered by groundwater flowing through cracks.

flow deposit. Unlike the muddy sandstone, the debris flows have clasts of many sizes. Each bed has a thin layer of sandstone about 1 inch (2.5 cm) thick at its top; these layers were deposited by smaller flows of muddy water that passed over the upper surface of each debris flow bed.

Two obvious characteristics of the conglomerate beds attest to the thick consistency and impressive strength of the debris flows: the upward-coarsening nature of each bed, and the size of some of the clasts. The beds range from marble-sized or golf ball–sized pebbles and finer-grained sediment at the base to bowling ball–sized clasts higher up. The biggest clast in the upper of the two beds is the size of a small microwave oven, and it protrudes through the thin layer of sandstone at the top of the flow deposit and into the base of the bed above. Upward-coarsening is common in debris flow deposits because of the manner in which the flows move. Most debris flows can be separated into a lower portion, which is sheared across the ground surface, and an upper portion, which moves along more or less as a single mass. Most of the vertical mixing within the flow takes place in its lower part, close to the ground. Larger clasts tend to be pushed

Two well-developed debris flow deposits, each about 6 feet (1.8 m) thick, are displayed above the man's head at Chimney Rock. The large angular boulder *(arrow)* protruding above the upper surface of the second deposit was rafted along in the powerful flow, clearly indicating that the flow had a very thick consistency.

upward, away from this zone of mixing. Once pushed into the upper part of the flow, they can be carried along like corks floating down a river. When a debris flow stops moving, its viscosity is high enough that large clasts are essentially frozen in place.

If you walk closer to Chimney Rock, you'll see that most of the clasts in the deposits are angular, and that virtually all of them are pieces of volcanic rock. These clasts traveled a relatively short distance—probably no more than 15 miles (24 km)—down the southern flank of Sunlight Volcano. They were not rounded much in their journey because the thick consistency of the flows cushioned the impacts between clasts. Their angularity and the fact that they are not sorted by size are characteristics of debris flow deposits. In contrast, river deposits tend to have rounded clasts that are sorted by size.

The debris flow deposits and minor sandstone making up Chimney Rock and the spectacular cliffs in the Shoshone River canyon also contain bits and pieces of petrified wood, along with smaller fossilized plant fragments that tend to be white because of bleaching caused by weathering. With careful searching you can find pieces of fossilized plant debris along the ground at Chimney Rock. However, a much better place to examine the famous petrified wood of the Absaroka Volcanic Supergroup is in the headwaters of the Tom Miner Basin, discussed in vignette 9.

Close-up of a debris flow deposit at Chimney Rock. The individual clasts are angular and poorly sorted. Pen for scale.

9.
Fossilized Forests
Petrified Wood in Tom Miner Basin

Among the most intriguing aspects of the sedimentary rocks belonging to the Absaroka Volcanic Supergroup is the fact that these rocks contain abundant petrified wood. This fossilized Eocene-age wood is noteworthy for its detailed level of preservation. Quite commonly, the internal structure of the fossil wood is so well preserved that tree rings are easily seen. You can also see individual cells with a hand lens or magnifying glass despite the tens of millions of years that have elapsed since the trees were living. Not only are the fossil trees well preserved, they are also diverse and include many subtropical varieties whose natural range does not extend anywhere close to Yellowstone today. Combined, the diversity and excellent state of preservation have yielded much important paleobotanical information that is useful for scientists trying to understand the landscapes and climate in North America during Eocene time. The diversity of tree types and presence of many subtropical varieties is particularly interesting, because during Eocene time the equator—in relation to Yellowstone Country—was located approximately where it is today. Unlike during Paleozoic time, when the Yellowstone region was located closer to the equator, the hotter climate in the Yellowstone region during Eocene time was caused by much higher global temperatures

As discussed in vignettes 7 and 8, rocks of the Absaroka Volcanic Supergroup were produced by a series of tall, majestic volcanoes that were active in Yellowstone Country roughly 55 to 48 million years ago. A green blanket of vegetation that included many large trees covered the steep upper flanks of the volcanoes as well as the alluvial apron that was deposited along their bases and across the valleys in between.

As with Yellowstone's forests today, the Eocene forests included conifers such as larch, spruce, pine, and hemlock—trees that grow at high elevations in regions characterized by strong seasons and seasonal limitations on the amount of available moisture. In addition to these trees, however, yews, cypresses, cedars, and an impressively diverse assemblage of flowering trees such as dogwood, persimmon, witch hazel, horse chestnut, laurel, magnolia, ash, sycamore, gum, and willow thrived along the flanks of the volcanoes at lower elevations and in the valleys between. These tree species require considerably more moisture and a climate that is warmer than that which occurs today in the northern Rocky Mountains.

Indeed, the climate was significantly warmer during Eocene time. Based on analyses of marine planktonic organisms, called *foraminifera*, preserved in layers of marine sediment of Eocene age, scientists estimate that ocean temperatures near the polar regions were about 9 degrees Fahrenheit (5 degrees C) warmer than today. Atmospheric temperatures in the same regions are estimated to have been about 18 degrees Fahrenheit (10 degrees C) warmer. At a global level, peak atmospheric temperatures were about 22 degrees Fahrenheit (12 degrees C) warmer than today. This warming, which lasted several million years, is called the Eocene thermal maximum. At its peak, there were no permanent polar ice caps. Rainy tropical environments extended as far north as about latitude 45, near present-day Portland, Oregon. Palm trees grew in the high Sierra Nevada in California, and huge inland lakes spread across the lower basins of Wyoming and Utah. And in the Yellowstone region, a mixture of subtropical and temperate trees grew across the mountainous volcanic terrain.

The warm Eocene climate was caused by a rapid accumulation of carbon dioxide gas in the atmosphere. Carbon dioxide gas is one of several so-called greenhouse gases that absorb and reemit infrared radiation from the sun, resulting in a buildup of heat in the atmosphere. As more carbon dioxide gas accumulates and more heat is retained, global temperatures warm.

There are two main ideas as to what caused the buildup of carbon dioxide gas. Some scientists think that the melting of natural gas clathrates at the very end of Paleocene time released a large volume of methane into the atmosphere. Gas clathrates are water-based crystalline

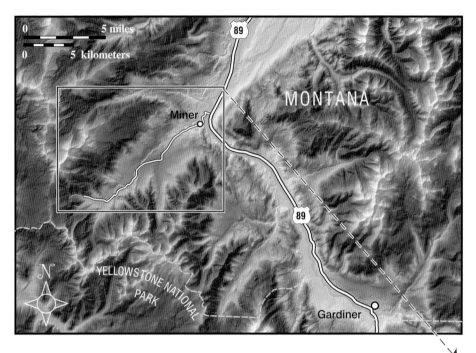

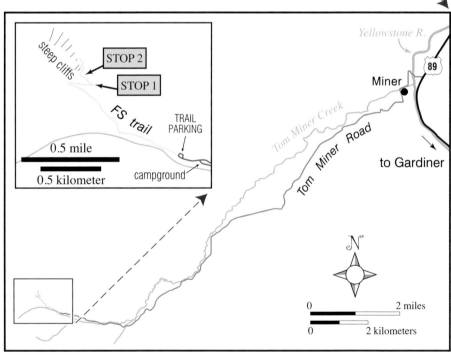

solids that physically resemble ice and contain molecules of natural gas, typically methane. Although plenty of mystery surrounds the formation of gas clathrates, they are widespread today in the marine sediment that is accumulating on the world's continental shelves, as well as in some onshore regions characterized by permafrost.

In this hypothesis, the methane that was released as the clathrates melted quickly broke down to carbon dioxide gas, causing both the atmosphere and the oceans to warm. It is estimated that global temperatures rose at least 11 degrees Fahrenheit (6 degrees C) in just 20,000 years. This temperature spike was followed by the Eocene thermal maximum, during which global temperatures rose more slowly. The increased temperatures would have caused additional clathrates

---

GETTING THERE

A short hike in the Tom Miner Basin, northwest of Gardiner, Montana, provides up-close views of debris flow deposits and the petrified wood preserved in them. The hike typically takes two to three hours at a leisurely pace and involves a climb of about 750 vertical feet (230 m) over about 0.5 mile (0.8 km). Be sure to wear sturdy footgear and bring extra clothing, food, and water.

To get there from Gardiner, follow US 89 north for 16.6 miles (26.7 km) and turn left (west) on Tom Miner Road at the sign for the Carbella Fishing Access Point. The turnoff is about 38 miles (61 km) south of Livingston, Montana. Follow Tom Miner Road to the bridge over the Yellowstone River and proceed 11.3 miles (18.2 km). The road is dirt and is not recommended for large recreational vehicles. After a number of twists and turns, the road ends at a turnaround next to a small U.S. Forest Service campground. Park at the turnaround and take the Gallatin Petrified Forest Interpretive Trail to the west. The trail climbs through the woods and switches back once before reaching stop 1—spectacular cliffs of debris flow deposits. After reaching the cliffs, the trail switches back again and continues upward and through the woods for about 150 yards (140 m) before coming to an obvious semicontinuous outcrop, which the trail continues along for another 100 yards (90 m). This outcrop is stop 2, and along it you can view several large petrified trees, tree roots, and smaller pieces of petrified wood. Although it's tempting to remove petrified wood from the area, it is illegal unless you have a permit from the U.S. Forest Service. If you do not have one, please leave all the petrified wood you find for others to enjoy.

to melt. Once this feedback loop had started, the process could not be reversed until essentially all the clathrates had melted.

The clathrate hypothesis regarding Eocene warming has gained significant support among scientists, although not everyone agrees. An alternative hypothesis is that that the burning of peat and coal caused a massive release of carbon dioxide gas into the atmosphere. Some scientists believe that the burning of Paleocene-age peat and coal deposits east and north of Yellowstone Country in eastern Montana and Wyoming was extensive enough to cause the global warming.

Regardless of the cause, it is clear that many of the tree species found in the Absaroka Volcanic Supergroup could only have existed if the climate was considerably warmer and wetter than today. Besides the presence of these species, the occurrence of tree rings in many of the woods indicates that, like modern trees, the growth rate of the trees sped up and slowed down in response to changing seasons. During

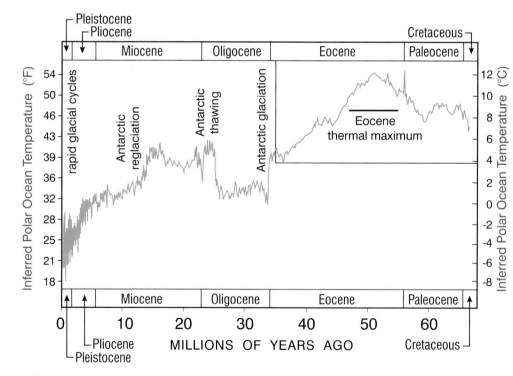

This graph shows how global temperatures fluctuated over the past 65 million years. Scientists constructed the graph by analyzing the change in the composition of small, fossilized foraminifera shells, which reflect the rise and fall of ocean temperatures. The warmest period was during Eocene time, when volcanoes were active in Yellowstone.

the period of the year most conducive to growth, the Eocene trees grew rapidly and produced earlywood, wood cells that are long with thin cell walls. In contrast, during the part of the year least favorable for tree growth latewood developed, wood cells that are narrow with thick walls. Each annual tree ring consists of earlywood and latewood. Although such seasonal variations can reflect changes in temperature from summer to winter, seasonal variations in moisture can produce the same effect even if the temperature does not vary much. For example, tree rings can form in regions with semitropical climates, where annual temperature fluctuations are minimal but there are significant changes in the amount of moisture available during rainy and dry seasons.

In many pieces of petrified wood from the supergroup, the annual tree rings can be seen with the unaided eye. The tiny rows of cells making up the earlywood and latewood can usually be seen with a magnifying glass or hand lens. The good preservation has long interested geologists and paleobotanists alike. Exactly how such delicate features in the original tissue become petrified—literally turned to stone—is an interesting question. The petrified wood is composed of quartz, a mineral composed of silica. Detailed studies using electron microscopes have revealed that the quartz occurs as tiny crystals, the smallest of which are only a few millionths of a millimeter (a few nanometers) long. The crystals are organized into patterns that reflect the cellular structure of the wood. In fact, the crystals that formed first actually used the organic molecules in the woody tissue as a template, thereby preserving the most delicate internal features of the original wood structure.

How exactly does organic woody tissue transform into the mineral quartz? Laboratory experiments show that silica-rich water, much like regular water, is capable of soaking into wood and causing it to become waterlogged. While this silica-rich solution seeps through the porous wood, the molecules of silica begin to link together, becoming longer chains and networks. These longer masses chemically bond with the wood, plating the organic molecules with thin layers of silica. After the first layer has been deposited, additional silica bonds to the newly laid substrate. Eventually, the original woody tissue itself is replaced by silica, and the entire mass of silica organizes into tiny crystals of quartz.

Most of the original woody tissue from trees, logs, and sticks buried in sediments of the Absaroka Volcanic Supergroup was replaced with silica through this molecule-by-molecule process. Silica was abundant in Yellowstone Country's groundwater during Eocene time because the region's volcanoes blanketed the landscape with huge quantities of silica-rich ash. The silica became partially dissolved in the groundwater.

Although scientists have known about the petrified wood in the Yellowstone region since the Hayden Expedition came across it in 1871, certain mysteries surrounding the wood have attracted the attention of paleobotanists and geologists during the last few decades. For example, within individual beds of conglomerate in the Absaroka Volcanic Supergroup, trees with distinctive annual rings (indicating a strongly seasonal climate) have been found next to trees with very faint rings (suggesting a climate with little seasonality). What's more, some conglomerate beds have been found to include pollen not related to any of

High-magnification image of petrified *Sequoia*—a cypress related to the giant redwoods growing today along the Pacific Coast—from the Absaroka Volcanic Supergroup. The transition from earlywood to latewood is clearly visible; the cells get smaller and the cell walls thicker across this transition.

the fossilized woods they contain. One bed that was studied contained an upright tulip tree trunk and seven upright gum tree stumps. The vertical nature of these stumps suggests they occur today where they grew originally. However, neither tulip nor gum pollen was found in the sedimentary rock matrix in which the trunks occur—a curious observation, because studies of modern trees have shown that much of the pollen a tree produces ends up accumulating in the sediment at its base. Instead, most of the plant pollen collected from the studied bed came from birch trees, although, interestingly, no petrified birch wood was found. All of these mysteries, at first puzzling, make sense when the sedimentology of these beds is considered. So let's start our hike and see what the rocks reveal.

The Gallatin Petrified Forest Interpretive Trail starts at the western end of the Forest Service campground. After cutting through sagebrush and aspens for about 0.5 mile (0.8 km), the trail weaves around several large boulders that have tumbled down through the trees from the steep cliffs to the north. The conglomerates and sandstones cropping out in the cliffs belong to the Washburn Group, part of the Absaroka Volcanic Supergroup. Each boulder consists of conglomerate with angular clasts of volcanic rock deposited by debris flows. The rocks of the cliff are part of the alluvial apron that blanketed the lower flanks of Eocene-age volcanoes located where the Gallatin Range is today, within about 5 miles (8 km) of the trail as the crow flies. When heavy precipitation events produced surface runoff, extremely powerful debris flows—mixtures of volcanic ash, water, and coarser debris—flowed down the volcanoes, ripping through forests and transporting soils, trees, and pollen from higher elevations to lower elevation regions where entirely different forest types grew. (See vignette 8 for more about the Eocene volcanoes and debris flows.) These deposits neatly explain the extremely diverse mixture of tree types and unexpected combinations of pollen and wood found in Yellowstone's fossil forests.

After passing the giant boulders, the trail begins to climb more steeply and switches back to the right. About 200 yards (180 m) past the switchback there is an impressive cliff rising straight above the trail. The lower, recessed third of the cliff is light-brown sandstone, whereas the upper two-thirds contains two spectacular overhanging sections of much darker brown conglomerate with very large angular boulders of

The dark rock with the large, angular blocks is a bed of conglomerate that was deposited by a single debris flow. The author's wife is standing next to a sandstone bed.

volcanic rock. The lower of the two overhanging sections consists of a single debris flow deposit, in which angular boulders the size of an overstuffed chair rest at the top of the flow. The base of this cliff (stop 1) offers a closer look at the angular clasts and jumbled organization that are characteristic of debris flow deposits.

There is a switchback just up the trail past stop 1, along which a small sign points out a hole in the outcrop. This quartz-lined hole is a geode. It is also a type of fossil called a *mold*, in which the impression of a plant or animal is preserved in rock as opposed to part, or all, of the actual fossilized organism. The hole formed as a log buried in the debris flow deposit decayed, and quartz crystals grew along the inner walls of the hole.

Continue walking along the trail, up and to the left as it climbs along the side of the hill. Be especially careful on this stretch of trail since it generally is more moist and slippery than the lower sections. After traversing a steep sidehill through the forest for about 150 yards

This mold just past stop 1 formed when a log buried in sediment decayed before it could be petrified. The white quartz crystals grew around the inside edges of the hole, partially filling it and forming a geode.

(140 m), the trail passes along the base of another cliff composed of debris flow deposits. The base of the cliff, which the trail continues along for about 100 yards (90 m), is stop 2. The cliff contains several well-preserved examples of petrified wood (marked with signs) as well as an impressively large boulder that was carried by one of the debris flows. The examples of petrified wood include several logs in which fossil tree rings can be seen without a hand lens or magnifying glass—in fact, without even leaving the trail.

Keep your eye out for small, whitish fragments of petrified wood that weathered out of the outcrop. They are scattered about on the ground at the base of the cliff. If you find a piece, chances are very high that you will be able to make out one or more tree rings. With a hand lens or magnifying glass you'll probably be able to see the groups of individual cells that make up the earlywood and latewood within the tree rings. As you examine the exquisite detail preserved in the petrified wood, keep in mind that these trees lived about 50 million years ago!

In more than one place along the cliff at stop 2 are large petrified tree trunks that occur in an upright position, suggesting that the trees are preserved where they originally grew. Many that occur in these Eocene deposits almost certainly were buried by debris flows where they grew; however, studies of modern debris flows, such as those that occurred after the 1980 eruption of Mount St. Helens, have shown that it is not uncommon for debris flows to uproot and transport stumps with their root wad and then deposit them pointing upward—that is, in growth position. Therefore, it seems at least possible that a similar process occurred with the debris flows that deposited much of the Absaroka Volcanic Supergroup. Some of the intact stumps at stop 2 may not actually have grown where they occur today, a possibility that fits right in with the strange combinations of tree types and pollen grains that have been found in these rocks.

Although the trail continues upslope and to the left for another 200 yards (185 m), it eventually dies out near the top of some steep

The upper half of the large boulder sticks out well above the top of the deposit of the debris flow that carried it, and its upright orientation indicates that the debris flow had a thick consistency—enough to keep the boulder from rolling over onto its side when the flow stopped moving.

outcrops. If you continue to scramble around, be sure to stay well away from the cliff edges, as there is plenty of "ball-bearing talus" on which you can lose your footing. Short bushwhacking forays above the end of the trail will take you to one or more large, upright petrified stumps sticking out of the sedimentary rock. In some of the gullies along the way, you might find large chunks of petrified wood that have weathered out of the cliffs. With a hand lens you'll be able to see tree rings and individual cells. More extensive bushwhacking in this part of the Yellowstone region will reveal many stunning examples of petrified wood, including vertical logs that cross more than one debris flow layer, a sign that the trees were buried by several flows while they were still standing. As you explore these rocks, please remember that—no matter how tempting—it is against the law to collect any petrified wood without a collecting permit from the U.S. Forest Service. Please leave everything you find for future generations to enjoy.

As you wander around this spectacularly beautiful part of Yellowstone Country, try to imagine the scene here about 50 million years ago. Impressive, very active volcanoes existed to the east and south. A steep, tree-covered alluvial apron was draped along their flanks and stretched across the intervening valleys. Following big precipitation

At stop 2 this petrified log, which is marked by a small sign, has distinctive tree rings.

events, powerful debris flows surged downslope, ripping trees in half, carrying portions of entire forests along, and mixing trees that grew high on the mountainsides with those that lived at lower elevations. Once the debris flows stopped, silica-rich groundwater soaked into the buried woody debris, petrifying it one molecule at a time.

The middle Eocene was the last time so many different subtropical tree species lived in Yellowstone Country. Following the Eocene thermal maximum, Earth's climate turned much colder, and by 34 million years ago the ice cap that still exists over Antarctica had started to form. Along with the extinguishing of Eocene volcanism, this trend toward a cooler climate drastically reshaped the Yellowstone landscape and profoundly changed the types of plants that grew across the region and the kinds of animals that could survive here. In vignette 10, we'll examine the geology associated with these landscapes and some of the now-fossilized animals that inhabited them.

A large petrified tree trunk in a gully beyond the end of the Petrified Forest Interpretive Trail in the Tom Miner Basin.

10.
From Playa Lakes to Rushing Rivers
Landscape Changes Recorded at Hepburn's Mesa

Visitors entering or leaving Yellowstone National Park through its North Entrance will drive through Paradise Valley—a wonderfully scenic piece of Montana stretching between Gardiner and Livingston. Along US 89 between these two towns are eye-catching, light-colored cliffs consisting mostly of siltstone. These cliffs, part of what's called Hepburn's Mesa, contain important clues regarding the landscapes that prevailed in the Yellowstone region after the Eocene volcanoes ceased to be active and the climate cooled (see vignettes 8 and 9).

After the Eocene volcanoes shut down, there was a period of about 27 million years during which little was preserved in Yellowstone's rock record. Either a long period of erosion took place or very little sediment accumulated from the rocks that were eroding. Sediment accumulation resumed in Yellowstone Country during Miocene time with the deposition of siltstone in one or more basins, at least one of which contained a playa lake at its center.

Playa lakes are temporary bodies of water that form in arid climates and in topographic basins that lack an outlet. The Great Salt Lake is a good example of a relatively large playa lake. As precipitation falling within the basin collects at its topographic low point, the playa lake grows. During relatively dry periods the lake shrinks. This shrinking and expanding has the effect of smoothing both the lake bottom and the adjacent onshore area, called the *lake plain*. Viewed together, the lake plain and lake bottom form a relatively flat, topographically featureless surface that slopes gently toward the center of the basin. The slope of most playa lakes is so gentle and the surface so flat that beachgoers commonly are able to wade hundreds of yards or more from the shoreline and still be in water that is only shin-deep!

At stop 1, Hepburn's Mesa is the hill with prominent cliffs of well-layered white to tan siltstone (and a minor amount of sandstone) about 0.25 mile (0.4 km) to the east across the Yellowstone River. The thickness of each individual layer stays about the same from one end of the outcrop to the other. These characteristics reflect the relatively flat, featureless topography associated with a playa lake and lake plain. This was a much different environment than the steep, thickly forested volcanoes that dominated the landscape during Eocene time. The volcanoes had been deeply weathered, and in the Hepburn's Mesa area, the rocks related to this volcanism had been buried by the fine-grained sediment making up the cliffs.

Some of the layers consist of volcanic ash that fell on the lake surface and accumulated on the lake bottom. The ash itself consists of silt- and sand-sized pieces of volcanic glass and tiny fragments of various minerals, most of which were derived from volcanic eruptions. Some of the eruptions came from the present-day shared borders of Oregon, Nevada, and Idaho, where the nascent Yellowstone Volcano (see vignette 11)

GETTING THERE

The light-colored siltstone cliffs of Hepburn's Mesa can be seen from a distance at stop 1, the intersection of US 89 and Big Creek Road, located 0.3 mile (0.5 km) north of a roadside rest area on the east side of US 89 between Gardiner and Livingston, Montana. The rest area is 23.7 miles (38 km) north of Gardiner and 29 miles (47 km) south of the I-90 exit in Livingston. If you have a set of binoculars, this is a good spot to use them. Stop 2 provides an up-close view of the cliffs. To get there from stop 1, drive south 4.5 miles (7.2 km) on US 89 and take the sharp left onto East River Road. Stop 2, the white cliffs of Hepburn's Mesa, are about 4 miles (6.4 km) north of the intersection. If you are coming from the north, you can bypass stop 1 to get to stop 2. About 22 miles (35 km) south of Livingston, turn left (east) onto Murphy Lane at the small settlement of Emigrant. Continue for about 1 mile (1.6 km) and turn right (west) onto East River Road. Hepburn's Mesa is 6.9 miles (11.1 km) south of the intersection. Park along East River Road and take a look at the base of the outcrop next to the road. Although you may be tempted, don't venture up onto the cliffs, as they are mostly located on private land and are very unstable.

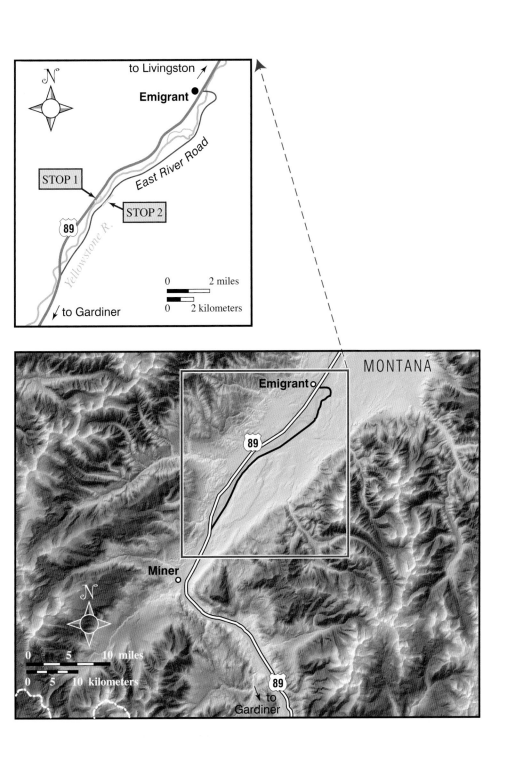

was beginning to erupt. Others were located more directly to the west in the Cascade Range of Oregon and Washington or to the southwest in the Great Basin of Nevada.

Other layers formed when the Miocene playa lake shrank and the Hepburn's Mesa region was transformed into a dry plain. During these times, soil developed on the exposed ground surface. Soils form when sediment and rock near the surface are altered by weathering, changed through the formation of various minerals, and affected by colonizing plants and animals and the accumulation of organic matter. As we'll see at stop 2, several characteristics related to ancient soil development are present in the light-colored sedimentary rocks at Hepburn's Mesa.

The cliffs at Hepburn's Mesa have yielded a very diverse and well-preserved assemblage of mammal fossils. These include extinct species of rabbits, squirrels, mice, moles, mountain beavers, deer, camels, antelope, and an early version of the modern horse. By comparing these fossils with those found in rocks of known ages elsewhere, paleontologists have determined that the light-colored cliffs are about 16 million years old at the base and about 14.8 million years old at the top.

The fossilized animals that have been discovered at Hepburn's Mesa evolved rapidly during a time in which a cooling climate altered environments. This cool period started as global temperatures began to decline following the Eocene thermal maximum (see vignette 9). What caused the global cooling is the topic of serious debate among climate scientists, but it probably had something to do with the removal of large quantities of the greenhouse gas carbon dioxide from the atmosphere. Some scientists have suggested that huge blooms of a small freshwater fern of the genus *Azolla* occurred in the Arctic Ocean at the end of Eocene time and removed enough carbon dioxide from the atmosphere to reverse the greenhouse effect. Regardless of its exact cause, the declining global temperatures and changing environments from Eocene through Oligocene and into Miocene time caused many different mammal species to evolve and then go extinct.

The climate also grew more arid than that which supported the lush forests of Eocene time. Although the presence of a playa lake itself suggests an arid climate, other clues are contained in the suite of fossil animals found at Hepburn's Mesa. For example, one type of fossil rodent, called **Mesoscalops**, had a skeletal structure that was specifically adapted for digging in the dry silty soils that prevailed in the gently sloping lake plain. This rodent had a long, scoop-shaped upper arm bone (humerus) and an exceptionally broad hand. The shape of the humerus allowed the lower arm to rotate as *Mesoscalops* scratched through the dirt with broad hands. As it pushed its hands downward and rotated them outward, the rodent used its unusually large neck and upper shoulder muscles to push its head upward like a lever, with its partially fused vertebrae serving as a fulcrum.

Although *Mesoscalops* has been compared to both extinct and modern true moles, the skeletal differences between *Mesoscalops* and true moles are so significant that paleontologists have concluded they do not even belong to the same family. For example, a modern true mole does not have fused vertebrae and does not use its head much for digging; rather, it uses its front arms like paddles to push dirt laterally away from its body without any rotation. Although *Mesoscalops* thrived in the dry, loose soils that formed during Miocene time in parts of Yellowstone Country, it became extinct shortly thereafter and has no known modern descendant.

Major changes in the fossils of larger mammals recovered from Hepburn's Mesa also reflect the transition to a more arid climate and, in particular, to an environment in which grasses dominated over leafy

The cliffs at Hepburn's Mesa as seen from stop 1. The cliffs are made of siltstone and a minor amount of sandstone and contain a famous suite of mammalian fossils. Lying above the white cliffs is a dark brown to gray conglomerate that has been interpreted as having been deposited by an ancestral version of the Yellowstone River. The black rock forming the skyline consists of two separate basaltic lava flows, although the two flows cannot be easily distinguished from this view.

vegetation. The horse genus *Merychippus* was the first to evolve molar teeth with high crowns and hard covering. Ancestors of *Merychippus* had teeth with short crowns and pointy, cone-shaped caps. Such teeth could efficiently pluck leafy vegetation but were not well suited for grinding it by chewing. The high-crowned teeth found in *Merychippus* wore down as they grew, and they had flat surfaces well adapted for grinding. Grasses contain microscopic bodies of solid silica called *phytoliths*. These bodies strengthen the structure of the grass but cause it to be more abrasive. With a tooth structure able to cope with the abundant and abrasive grasses, grazers like *Merychippus* were better able to survive.

Along with changes in *Merychippus's* tooth structure, the middle toe of its foot developed into a hoof with ligaments that supported the horse's weight, replacing the padded foot of its ancestors. This change allowed *Merychippus* to run fast. At the same time, an increased skull size and, correspondingly, a larger brain resulted in a smarter, more agile animal better equipped to interact or communicate with other individuals and thereby escape predation. Relative to its slower, more docile ancestors, which lived in the forests, *Merychippus* was better adapted for traveling quickly across open grasslands in herds and was the first of the so-called speedy grazers—those animals specifically adapted to life on the open plains.

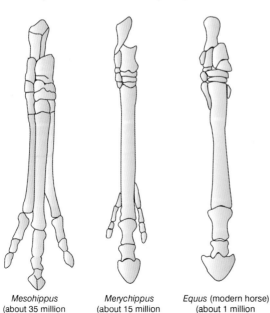

Mesohippus
(about 35 million
years ago)

Merychippus
(about 15 million
years ago)

Equus (modern horse)
(about 1 million
years ago)

Front leg bone structure of *Mesohippus* (an ancestor of *Merychippus*), *Merychippus,* and a modern horse. Relative to earlier horses, the two side toes on the foot of *Merychippus* were less well developed, but the middle toe was bigger and more stable, with a larger nail. These adaptations allowed *Merychippus* to gallop much like a modern horse, making it the first of the so-called speedy grazers.

One of the noteworthy attributes of the mammal fossils collected from Hepburn's Mesa is that many are intact entire skeletons, rather than individual bones or bone fragments. The exceptional quality of the fossil remains results from the fact that animals that died along the lakeshore or in the lake mud had a good chance of being preserved as whole skeletons, particularly if the lake expanded and covered the carcass shortly after the animal died. Unlike in open grassland environments, where carcasses are almost always scavenged and eventually undergo complete decay, animals that meet their demise in or near lakes stand a better chance of being preserved. Some paleontologists think that many of the smaller rodent skeletons, such as *Mesoscalops*, collected from Hepburn's Mesa were hairballs—undigestible bones and fur—regurgitated from birds of prey. Hairballs that ended up in the lake would have become waterlogged, sinking to the bottom where they were buried in the fine lake sediments. This quick burial preserved even the most delicate bones.

If you would like to see the white cliffs of Hepburn's Mesa up close, proceed to stop 2. Please exercise caution at stop 2 because the siltstone cliffs are very unstable and dangerous. Generally, the best approach is to examine some of the bigger blocks of rock that have weathered out of the cliff and rolled or fallen down to the road. I do not recommend climbing around on the steep, unstable hillside.

The first thing you will probably notice about the exposure is that fresh rock is hard to find. Most of the cliff is covered with a thin veneer of dried mud that looks like popcorn. In fact, geologists refer to this style of weathering as "popcorn weathering." Popcorn weathering reflects the presence of expandable clay minerals and is very typical of fine-grained sedimentary rocks containing significant amounts of volcanic glass. Much of the glass has weathered to clay. When the clay is wetted, it absorbs some of the water, causing the sediment to physically expand. When it dries out, the sediment shrinks and cracks. This physical expansion and contraction of the outer part of the outcrop causes it to break apart. When heavy enough rain falls, some of the loosened sediment washes down the steeper faces of the exposure, creating the mud veneer. This veneer undergoes expansion and contraction too, forming the irregularly shaped particles that resemble popcorn.

Close-up of popcorn weathering at stop 2. This texture is produced by the episodic wetting and drying of expandable clay within the siltstone. Rock hammer for scale.

Why does volcanic glass weather to clay? Glass forms in large, violent eruptions in which magma cools almost immediately. The glass is fragmented by the violence of the eruption into sand- and smaller-sized pieces, called *shards*, that are hurled skyward as part of a large plume of ash. Eventually, the glass shards settle to the ground, though sometimes not before the wind carries them thousands of miles. Volcanic glass lacks an organized crystalline structure because it cools so quickly; the atoms in the magma are frozen in place before they have time to organize into different mineral crystals. Since it lacks a crystalline structure, glass is not chemically stable. Over geologic time the randomly arranged atoms slowly organize to form a variety of clay minerals.

X-ray analysis has shown that the finest-grained samples from Hepburn's Mesa are made mostly of a zeolite mineral called *clinoptilolite*. The zeolite mineral group is popular with collectors because many of them are rare and have beautiful, delicate crystal shapes. Zeolites often form in ash-rich, fine-grained sedimentary rocks, and many are useful as industrial filters because their molecular structures have relatively large open spaces with consistent sizes and shapes. The dimensions of the vacant spaces vary with each zeolite mineral, but some of the more useful zeolites, such as clinoptilolite, are used as molecular sieves— superfine filters that trap and retain other molecules of a specific size and shape.

Scanning electron microscope image of clinoptilolite crystals from Nevada that are similar to those found in the white cliffs of Hepburn's Mesa. The small bar at the bottom left is 0.5 millimeter long. —Courtesy of International Natural Zeolite Association

Although clinoptilolite has never been mined at Hepburn's Mesa, it is mined from playa lake deposits in the southwestern United States and is used in a variety of applications, including as a molecular sieve for various chemicals, a gas absorber, an odor control agent, and a water filter for municipal and residential drinking water and aquariums. The fact that clinoptilolite performs well over an extreme range of temperatures and is chemically neutral adds to its value as an industrial resource. What might strike you as surprising is that for several years now clinoptilolite has been used as a feed additive for cows, pigs, horses, and chickens. It absorbs toxins in the feed that are created by molds and microscopic pathogens, and it appears to enhance food absorption by these animals. Similar uses in human food are currently being tested.

The clinoptilolite at Hepburn's Mesa formed as volcanic glass in the silty sediment was chemically altered in the presence of saline water. Because saline water is common in arid climates, particularly in playa lakes, the abundance of clinoptilolite at Hepburn's Mesa provides supporting evidence that Yellowstone Country experienced an arid climate during middle Miocene time. The strange, bloblike, yellowish brown objects that range from 1 to 6 inches (2.5 to 15 cm) across and stick out of the cliffs provide additional evidence. These objects may look like fossil bones, but they are actually nodules of the mineral calcite.

The calcite nodules reflect variations in the amount of water reaching the center of the basin through time—variations caused by differences in the amount of precipitation and the intensity of evaporation over annual and longer time frames. Playa lakes are famous for their rapidly expanding and shrinking shorelines, and during droughts they can dry up altogether. Although there is no water at the surface in the dry basin, water still exists underground below the water table. The water table marks the place underground below which the pores between sediment particles are filled with water. Above the water table the pores are filled mostly with air. Water is drawn upward from the water table and wets the sides of the pores that are otherwise filled with air. A similar thing happens when a dry sponge is placed in a puddle of water on a countertop: the water slowly soaks upward into the sponge. Eventually, most of its holes will contain some water and some air.

During Miocene time, the periodic wetting of the pores by saline groundwater in the shallow subsurface introduced calcium and carbonate in dissolved form, whereas subsequent drying of the pores caused calcium and carbonate to be precipitated as calcite. Commonly, precipitation is concentrated in and around fossil burrows or root traces (hollows left after roots have decayed) in sediment that provide natural

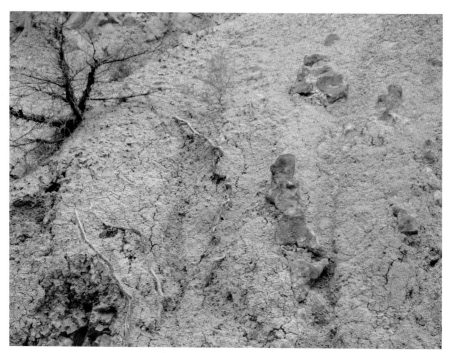

Brownish calcite nodules in the cliffs of Hepburn's Mesa. The small shrub in the upper left corner is about 18 inches (46 cm) tall.

pathways for saline groundwater to be wicked upward. Calcite nodules can also be precipitated randomly within sediment, rather than being associated with preexisting features.

There are other burrows in the cliffs that are not associated with calcite nodules. Most of the burrows are about 0.5 inch (1.3 cm) or so in diameter and are oriented vertically in the sediment. Based on comparisons with modern burrows in similar environments, these were probably produced by large insects such as cicadas, which tunneled into the dry soils of the lake plain.

Rounded cobbles and boulders are common around the base of the cliffs. These rounded clasts weathered out of the conglomerate layer resting on top of the white cliffs. Many of the clasts are covered by small circular marks ranging from about 0.5 to 1 inch (1.3 to 2.5 cm) in diameter. Called *percussion marks*, these circular features formed as the clasts were transported in a river and had forceful collisions with other large clasts. Together with the well-rounded shape of the cobbles and boulders, the percussion marks attest to the power of the ancient river that transported these clasts prior to depositing them here.

A small burrow in white siltstone from Hepburn's Mesa. A large tunneling insect such as a cicada probably produced this burrow, which was then filled in with silt.

Geologists studying the conglomerate layer have reported that the clasts within it are arranged in a domino-type fashion, in which individual clasts are touching one another and are tilted at a slight angle. Called *imbrication*, this arrangement forms as flowing water moves clasts and arranges them in a stable position. Geologists can determine the direction the flow went based on the orientation of the clasts it deposited. In this case, the clasts were deposited by a river that flowed north, the same direction as the modern Yellowstone River.

The river likely was an ancestral version of the Yellowstone River that flowed through the Hepburn's Mesa area between 14.8 and 8.6 million years ago, the former being the approximate age of the top of the white cliffs and the latter the age of the oldest basalt overlying the conglomerate. Viewed from above, the basalt flows forming the upper layer of rock at Hepburn's Mesa are elongate in a southwest-northeast direction, indicating that they flowed down the ancestral Yellowstone River valley. Although the flows probably altered the course of the river

The rounded clasts composing the conglomerate at the top of Hepburn's Mesa were likely transported and deposited by an ancestral version of the Yellowstone River.

somewhat, they did not completely fill the valley, and the Yellowstone River has continued to flow northward down this portion of its valley for at least the past 8.6 million years.

What caused the landscape of Hepburn's Mesa to change so drastically, from an open, grassy basin with a playa lake at the center to an elongate, northward-trending valley with a river rushing through it? The basalt capping Hepburn's Mesa is a big clue. Although it's not obvious from either stop 1 or stop 2, the basalt actually consists of two separate flows, one 8.6 and the other 5.2 million years old. Geologists think both are related to a volcanic center that was slowly drifting northeastward toward Yellowstone from the center's birthplace in the remote country where the borders of Oregon, Nevada, and Idaho meet. Although the volcanic center did not reach Yellowstone Country until about 2.1 million years ago, heat associated with the center caused the rocks along its leading edge to expand in volume, lifting the land surface prior to the arrival of the center itself. The conglomerate, composed of clasts shed from the uplifted land surface, is the first indication that significant topographic relief had developed south of Hepburn's Mesa; it represents the first sign of landscape change associated with the arrival of Yellowstone's world-class volcano—the heart of the volcanic center. In vignette 11 we'll examine the geology associated with the first of three gigantic eruptions that have come from this volcano since it arrived in Yellowstone Country.

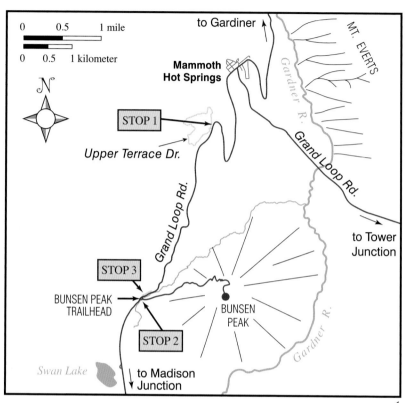

11.
Arrival of the Hot Spot
The First Caldera-Forming Eruption in Yellowstone

Yellowstone National Park sits directly on top of a large, very active volcano. Although the large size and subtle form of the volcano can make it difficult to appreciate, it has attracted the close attention of scientists since F. V. Hayden—leader of the Geological and Geographical Survey of the Territories—first described some of its elements in his official

---

GETTING THERE
Stop 1 provides a superb view of the Huckleberry Ridge Tuff, which was produced during the largest and oldest of Yellowstone's three caldera-forming eruptions. To get there from the Mammoth Hot Springs Hotel, drive south for 2 miles (3.2 km) on Grand Loop Road and turn right (west) onto Upper Terrace Drive (32.2 miles, or 52 km, north of Madison Junction). Proceed about 0.2 mile (0.3 km) to the small parking lot on both sides of the road. From the parking lot, look east toward Mt. Everts. For a closer look at the tuff, proceed south on Grand Loop Road for 2.7 miles (4.3 km) to the Bunsen Peak Trailhead (29.5 miles, or 47.5 km, north of Madison Junction). Walk about 0.25 mile (0.4 km) up the trail to stop 2, the fantastic overview of the tuff exposed at Golden Gate, about 0.25 mile (0.4 km) to the northwest. To get your hands on the Huckleberry Ridge Tuff seen at a distance from stop 2, continue to stop 3, the parking area for Rustic Falls, located at Golden Gate about 200 yards (180 m) north of the Bunsen Peak Trailhead on Grand Loop Road. Park at Rustic Falls, then proceed carefully across the road to the roadcut.

---

report to U.S. Congress in 1872, the year Yellowstone was designated as the world's first national park. Wrote Hayden:

> From the summit of Mount Washburn, a bird's-eye view of the entire basin may be obtained, with the mountains surrounding it on every side without any apparent break in the rim. It is probable that during the Pliocene period the entire country drained by the sources of the Yellowstone and Columbia [rivers] was the scene of as great volcanic activity as that of any portion of the globe. It might be called one vast crater, made up of thousands of smaller volcanic vents and fissures out of which the fluid interior of the earth, fragments of rock, and volcanic dust were poured in unlimited quantities.

Since Hayden's remarkably insightful report, many scientists have studied the Yellowstone Volcano and the geology that surrounds it. In 2001, the Yellowstone Volcano Observatory was established to monitor the entire Yellowstone region for signs of an eruption or other geological hazard. Frequent news coverage of earthquake swarms (multiple small earthquakes that occur in the same general region over time frames ranging from days to months) and other geologic activity associated with the Yellowstone Volcano has helped bring public awareness to the potential hazards it poses. Colorful TV documentaries, including the 2005 docudrama *Supervolcano*, have provided insights into the catastrophe that could result from another major eruption.

Stretching across nearly the entire park, the Yellowstone Volcano has a broader, more subtle form and is considerably larger than many of Earth's more classically shaped volcanoes, such as Mt. Fuji in Japan or Mt. Rainier in Washington. Most of the central portion of the park is located within a central crater called a *caldera*, which is Spanish for "cauldron." Calderas form when the ground collapses after a particularly big eruption. Since the caldera formed, several smaller eruptions have occurred within or along its margins (see vignette 13), though these more recent eruptions have not resulted in a caldera. At its widest, the caldera is about 40 miles by 30 miles (64 km by 48 km), and its approximate center is the Old Faithful Geyser area.

The Yellowstone Volcano is the modern-day manifestation of a volcanic center that has been active for about 16.5 million years. This center has produced dozens of caldera-forming volcanoes, each located in a slightly different place. The first occurred in the remote country where the borders of Oregon, Idaho, and Nevada meet. Since those earliest eruptions, the volcanic center has drifted from its birthplace to its present location in Yellowstone at a speed of a little less than 2 inches (5 cm) per year.

Although the leading edge of the center produced some earlier and relatively minor volcanism in the Yellowstone region, including basalt flows 8.6 and 5.2 million years ago that are exposed at Hepburn's Mesa (see vignette 10), the volcanic center itself did not arrive until about 2.1 million years ago. Since then, three caldera-forming eruptions have occurred. The first and largest eruption, about 2.1 million years ago, left behind a humongous caldera with an area of about 6,000 square miles (15,500 square km)—larger than the state of Connecticut! The remnants of this caldera span approximately 60 miles (97 km), from Canyon Village southwest to a point outside the park and well into

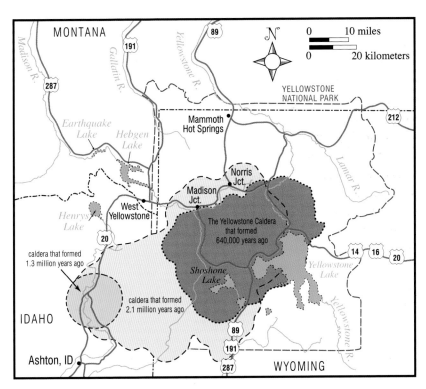

The Yellowstone Volcano has produced three separate calderas. Each formed as the central region of the volcano collapsed following a large eruption or series of smaller, closely timed eruptions.

Idaho. At its widest, the caldera was about 40 miles (64 km) across. The youngest caldera formed 640,000 years ago and has an area of about 4,700 square miles (12,170 square km)—still bigger than the states of Rhode Island and Delaware combined. We'll examine this caldera and the eruption associated with it in vignette 12. The smallest of the three calderas, called the Island Park Caldera, formed about 1.3 million years ago southwest of the park and north of the town of Ashton, Idaho. It is about eight times the size of the District of Columbia.

A column of exceptionally hot, partially molten rock is slowly rising from the mantle beneath the Yellowstone Volcano. The heat associated with this hot rock is called a *thermal plume*. When a thermal plume reaches Earth's surface, it produces a hot area called a *hot spot*. Dozens of hot spots on Earth have been studied, and many of these appear to have thermal plumes that extend downward through the lithosphere into the underlying mantle. For many years geologists

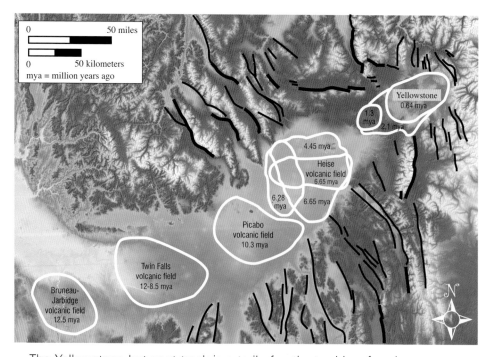

The Yellowstone hot spot track is a trail of extinct caldera-forming volcanoes in the eastern Snake River Plain of southern Idaho. The calderas become younger as they get closer to Yellowstone—the site of the currently active volcano. Many scientists regard the hot spot track as reflecting the southwestward drift of the North American Plate over a relatively stationary thermal plume.

postulated that the deepest plumes extend all the way to the core-mantle boundary, where they were thought to originate from the very hot surface of the outer core. Many geologists have regarded these deep thermal plumes as being fixed in place relative to the lithospheric plates that pass overhead.

Indeed, many published interpretations of the Yellowstone hot spot regard it as the near-surface expression of one of these deep stationary plumes, heating the North American Plate as it passes overhead. According to this idea, the movement has produced a trail of extinct caldera-forming volcanoes called the Yellowstone hot spot track. The hot spot track extends south and west of Yellowstone, which is located directly over the plume now, down the length of the eastern Snake River Plain. The track parallels the southwesterly direction that the North American Plate moves. The volcanoes along this track grow older the farther you get from the Yellowstone Volcano. Most of these extinct volcanoes have largely been buried by basalt flows.

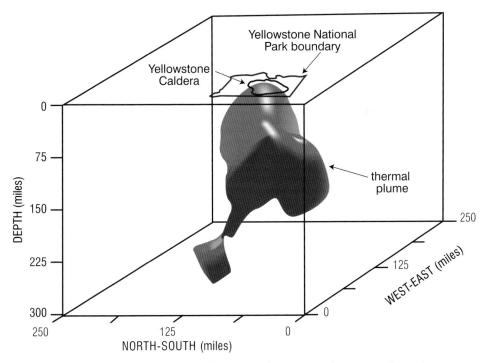

By measuring the velocity of earthquake waves that pass through Earth, seismologists created an image of the thermal plume beneath Yellowstone National Park. Earthquake waves slow down as they pass through the hot and partially molten rock of the plume.

The linear Yellowstone hot spot track and the recognition that the extinct volcanoes become progressively older to the southwest, away from Yellowstone, is seemingly compelling evidence in favor of the stationary thermal plume idea; however, this idea is being seriously questioned by many scientists. Some scientists are so convinced that the stationary thermal plume theory is wrong that they refer to it as a "scientific zombie"—an idea that is dead but simply won't go away.

Geophysical investigations indicate that the plume beneath Yellowstone is a rather shallow feature that extends downward only about 300 miles (485 km) below the surface into the uppermost mantle—far short of the core-mantle boundary, which lies about 1,780 miles (2,870 km) down. Furthermore, some hot spot tracks do not appear to parallel the direction overlying lithospheric plates have traveled. The Hawaiian Island chain, long regarded as a good example of the stationary thermal plume theory, is such a hot spot track. These observations suggest that thermal plumes are not deep stationary features over which lithospheric plates move.

Some scientists propose that the Yellowstone hot spot formed as a result of tectonic changes that occurred about 20 million years ago. This theory is quite technical, but it posits that a massive fracture developed in the lithosphere as a result of a spreading center being subducted beneath North America. The fracture in the lithosphere propagated toward Yellowstone, moving in a northeasterly direction. The thermal plume and overlying hot spot followed the tip of this propagating fracture, forming the hot spot track.

Although the scientific debate about the origins and mechanics of the Yellowstone hot spot no doubt will rage on for years to come, geologists agree that the three caldera-forming eruptions of the Yellowstone Volcano were immense—far bigger than any eruption in recorded history. Not only were the craters left behind huge, but the amount of material erupted was staggering. The total volume of volcanic ash spewed out during the eruption 2.1 million years ago is estimated to have been more than 600 cubic miles (2,500 cubic km), the equivalent of a cube of rock that measures almost 8.5 miles (13.7 km) on a side! This was about 2,500 times the volume of ash produced during the 1980 Mount St. Helens eruption in Washington—an eruption that killed fifty-nine people, leveled mature forests for miles around, and deposited as much as 1.5 inches (4 cm) of grit up to 180 miles (300 km) away in eastern Washington and surrounding regions. By comparison, ash erupted from the Yellowstone Volcano blanketed most of western North America, reaching more than 300 miles (480 km) out to sea off the coasts of Oregon and northern California and eastward nearly to the Mississippi River, which dumped ash into the Gulf of Mexico. In the immediate Yellowstone region, the ash exceeded 600 feet (185 m) in thickness. If all of the ash from the eruption was spread evenly across California, it would be nearly 20 feet (6 m) thick. The same volume spread across New York State would be more than 58 feet (17.7 m) thick, and across the District of Columbia it would be nearly 9 miles (14.5 km) thick!

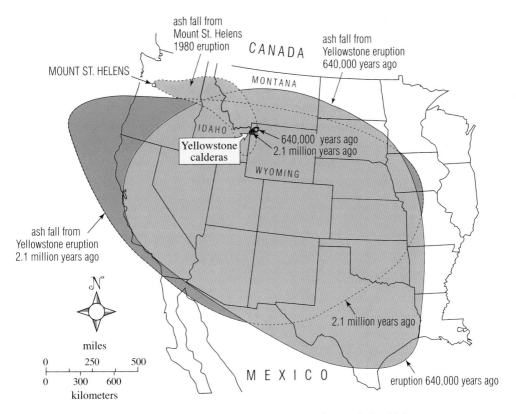

Distribution of volcanic ash from the eruptions of the Yellowstone Volcano that occurred 2.1 million years ago and 640,000 years ago, both of which were bigger than any in recorded human history. The distribution of ash from the 1980 Mount St. Helens eruption is shown for comparison.

The caldera-forming eruption that took place 2.1 million years ago actually consisted of three separate, smaller eruptions from different parts of the Yellowstone Volcano. Even so, each eruption was tremendously large in size and produced gargantuan quantities of volcanic material.

Each eruption probably began with highly explosive outbursts of fine ash, bits of magma, and pieces of rock torn from the walls of an opening called a *vent*, which formed in the crust within the volcano. Ash probably left the vent at supersonic speed—faster than the speed of sound. The booming from these explosions would have been heard hundreds of miles away. The ash cloud likely was blown into the atmosphere up to 50,000 feet (15 km)—heights measured during modern highly explosive eruptions. In the immediate Yellowstone region, this ash was so thick that it would have blocked out the sun, plunging the region into temporary darkness. Although some of the ash fell back to Earth over the days that followed the eruption, the finest ash would have remained in the atmosphere for a year or more, carried around the world by the jet stream and resulting in beautiful sunsets throughout much of the northern hemisphere.

As the eruption continued, tremendous volumes of ash, gas, and rock were hurled skyward, forming an eruptive column over the vent, while powerful earthquakes rocked the region. Close to the ground, the column collapsed as most of the ejected material fell back to Earth around the still-erupting vent. The returning deluge spread out laterally, forming turbulent flows of hot ash, gas, and rock called *pyroclastic flows*. These flows raced across the landscape at speeds in excess of 100 miles per hour (160 km per hour), following topographically low regions like river valleys. Pyroclastic flows are extremely dangerous and can be so hot that they are incandescent. Usually nothing can survive in their path.

Both the ash and debris from the initial eruptive blasts and the powerful pyroclastic flows that followed produced a rock type called *tuff*. The deposits from an initial explosive blast commonly form what is called *ash-fall tuff*, which consists of a loosely packed jumble of small pieces of volcanic rock and ash; ash-fall tuff is usually well layered. In contrast, pyroclastic flows deposit what is called *ash-flow tuff*. Ash-flow tuffs can contain enough heat that they partially melt after being deposited, causing the accumulated material to continue to deform plastically. As an eruption continues, additional pyroclastic flows add more hot material to the top of the growing volcanic deposit. This additional weight causes the lower parts of the deposit, which contain the most residual heat, to compact and fuse together, forming a volcanic rock called *welded tuff*. The entire deposit produced by a pyroclastic flow is called an *ignimbrite*, from the Latin *igni* (fire) and *imbri* (rain).

The oldest caldera-forming eruption of the Yellowstone Volcano produced three separate ignimbrites. Based on mapping done across Yellowstone Country, geologists have determined that each was derived from a separate segment of the volcano. These three ignimbrites make up Members A, B, and C of the Huckleberry Ridge Tuff. Because the ignimbrites do not grade laterally into each other but rather occur one on top of the other, it is clear that the three eruptions were not simultaneous but were separated by an interval of time, as little as a few hours or as much as several years.

At stop 1, look eastward across the Gardner River valley to the summit of Mt. Everts—the high ridge on the skyline. A prominent layer of reddish brown rock forms many of the highest outcrops on this ridge. The upper half of the reddish brown outcrop belongs to Member B, whereas the lower half belongs to Member A. Member A, in turn, rests on top of layered, gray and yellowish Cretaceous-age shales and sandstones of the Everts Formation. The surface separating the Everts Formation from the base of Member A is an unconformity that represents about 63 million years—the time that elapsed between the deposition of the Cretaceous sedimentary rocks about 65 million years ago and the eruption of the tuff about 2.1 million years ago.

In places, the uppermost 20 feet (6 m) or so of the Everts Formation has a distinctly red color. When the hot tuff was deposited, it heated the ground surface and baked the underlying rocks, causing red iron oxide minerals to form. Those with good eyes or binoculars will see a thin whitish pink layer—only about 8 feet (2.4 m) thick—at the bottom of Member A. This is ash-fall tuff, which represents the initial release of pressure from the volcano's magma chamber and the first phase of the eruption. You'll notice that the ash-fall tuff deposit is quite thin relative to the overlying cliff of Members A and B of the Huckleberry

Ridge Tuff, which was deposited by two huge pyroclastic flows that followed the initial blast.

A second overview of Members A and B of the Huckleberry Ridge Tuff can be had at stop 2, where their overall features can be more easily appreciated from a distance of about 0.25 mile (0.4 km). Look northwest, across the canyon containing Glen Creek, to the prominent layers of purple and brown rock outcropping along the far side of Grand Loop Road. The canyon wall consists of steep cliffs separated by a gentler slope. The lower cliff, Member A, is purplish brown and consists of one main, massive layer cut by numerous vertical cracks called *columnar joints*. The top of Member A contains a few crude layers, each about 5 to 10 feet (1.5 to 3 m) thick. Member B occurs above the slope that separates the two cliffs and is much better layered, with individual layers ranging in thickness from less than 1 foot (30 cm) to more than 5 feet (1.5 m).

Looking east from stop 1 toward Mt. Everts and the Huckleberry Ridge Tuff, which rests on top of Cretaceous-age sedimentary rocks. The thin orange layer directly below the tuff consists of sedimentary rocks that were baked by the hot pyroclastic flows.

The differences in layering between the two ignimbrites reflect how far the pyroclastic flows traveled from their vents and the intensity of the eruptions that created them. Most pyroclastic flows are actually a series of individual surges separated by minutes or days. Ash that accumulates close to the vent it was erupted from is likely to be hotter and more easily welded than ash that has traveled farther in a surge. Because the vent from which Member A erupted was relatively close

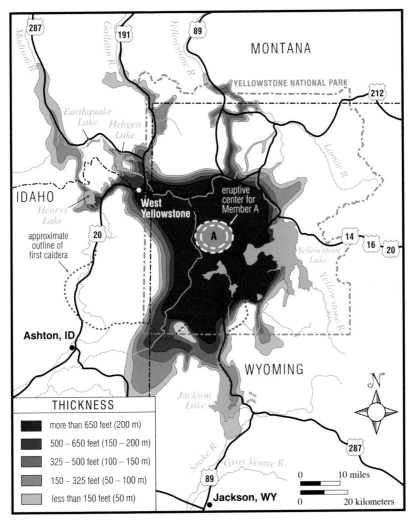

The thickness of Member A of the Huckleberry Ridge Tuff. Long fingers of tuff accumulated in the Madison, Gallatin, and Yellowstone river valleys, indicating that the valleys must have been present 2.1 million years ago when the pyroclastic flows occurred.

to stop 2, most of the member exposed at stop 2 was deposited by a semicontinuous pyroclastic flow. The material was hot, thick, and dense enough that it fused together as one big mass called a *cooling unit*. Shortly after it accumulated, the mass of the overlying deposit flattened bubbles and other pore spaces within the ash-flow tuff, while small bits of taffylike magma cooled to glass, transforming the entire member into a welded tuff. The columnar joints developed as the whole mass contracted and cooled as a single unit.

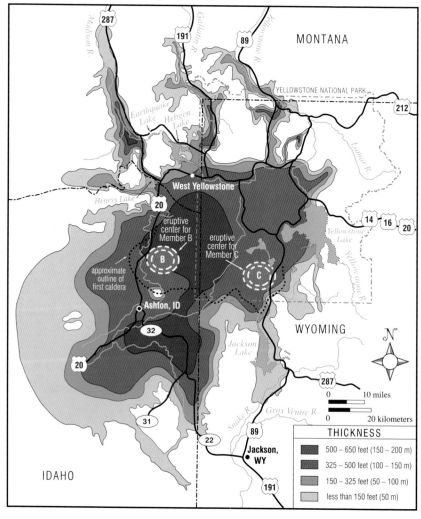

The thickness of Member B of the Huckleberry Ridge Tuff. The vent for Member B was farther away from stop 1 than that which produced Member A.

In contrast, the vent from which Member B erupted was farther away from stop 2. Although crude columnar jointing in the thicker layers of tuff at the base of Member B indicate that the initial surges formed a single cooling unit, most of Member B is well layered. The layering indicates that the surges had traveled far enough from their vent that they were more clearly separated in time, more diffuse, and probably a bit cooler when they came to rest here. Most of the layers in the upper part of Member B represent individual cooling units.

You can take a close look at Member B at stop 3, where the tuff is exposed along the road. If you walk to the northern end of the accessible roadcut, you can also see the uppermost part of Member A. The contact between the two members corresponds to the small topographic bench separating the lower cliffs from the upper cliffs. In Member B, relatively large crystals of white feldspar and clear quartz, all about $\frac{1}{8}$ inch (3 mm) or so across, are encased in a pinkish matrix. The feldspar crystals are made of the mineral sanidine and typically have a blocky shape. Both minerals were growing in the magma chamber below the surface before being spewed out of the volcano during the eruption.

View of the Huckleberry Ridge Tuff from stop 2. The lower cliff with the crude columnar jointing is Member A; the upper cliff with better-developed horizontal layering is Member B.

The fine-grained matrix of the rock is made of volcanic ash and small bits of volcanic glass that formed as the magma cooled rapidly.

As stated before, Members A, B, and C were erupted from three different parts of the Yellowstone Volcano. Each member tapers away from the vent it erupted from. (I have used the singular *vent* for simplicity, but it's possible that each member was erupted from multiple vents clustered in a region.) Relatively large crystals, such as those seen in Member B at stop 3, dropped out of the pyroclastic flows close to the vent from which they came. Although all three vents have been buried or mostly buried by volcanic rocks, geologists have been able to estimate the location of the vents by mapping the abundance of these crystals and comparing this information to the shape and thickness of each member. The vent that produced Member A was located in the vicinity of Upper Geyser Basin, which today contains Old Faithful Geyser. The vent that produced Member B was southwest of Yellowstone National Park, near the town of Ashton, Idaho. Member C erupted from the south-central corner of the park.

Close-up of Member B at stop 3. Clear quartz crystals occur in a finer-grained pink matrix of ash and volcanic glass. Pen tip for scale.

Among the more distinctive features of both members exposed at stop 3 are horizontally elongated objects that are about 1 inch (2.5 cm) or more long and recessed into the outcrop. Most of the objects are fragments of pumice, a very low-density volcanic glass. Pumice forms as frothy magma full of dissolved gas cools rapidly. The gas bubbles are frozen in the foamy magma as it cools, giving the pumice its porous texture. These bits of pumice were incorporated into the pyroclastic flow that formed both members. After the flow came to rest, the taffylike fragments were compacted as their bubble holes collapsed, resulting in the horizontally elongated features.

Imagine being able to watch Yellowstone's first caldera-forming eruption 2.1 million years ago. As magma and gas in the shallow parts of the magma chamber began to erupt from a vent near the present-day Upper Geyser Basin, powerful earthquakes rocked the region. The initial blast sent a plume of hot ash skyward, some of which fell back to Earth as the ash-fall tuff layer seen at stop 1. As the eruption increased in intensity, a gargantuan mass of hot gas, magma, and ash erupted,

Close-up of partially collapsed pumice fragments near the base of Member B at stop 3. The relatively soft pumice has weathered more easily out of the rock, leaving these holes. Rock hammer for scale.

forming a giant column that soon began to collapse. Powerful pyroclastic flows spread out laterally from the collapsing column. These flows raced down each major drainage system near the vent.

The pyroclastic flow that formed Member A of the Huckleberry Ridge Tuff roared down the Gardner River valley toward Mt. Everts, leveling trees and burying the valley with more than 300 feet (90 m) of searing volcanic debris. Incandescent gases in the flow gave it an ominous, hellish look. Any living thing caught in its path was incinerated. The flow continued downstream through the Yellowstone River valley, reaching past the town of Gardiner. At the same time, similar large pyroclastic flows rushed down the Gallatin and Madison river valleys to the northwest.

Perhaps a few days—or years—later one or more vents southwest of Yellowstone National Park began to erupt, sending pyroclastic flows down the Gardner River valley yet again and into the Yellowstone River valley. Most if not all of these flow surges also were very hot, and collectively they formed Member B. These flows were able to travel farther down the Yellowstone River valley because the pyroclastic flows preceding them had flattened forests and buried small hills, thereby smoothing over the landscape. Some surges associated with Member B even spilled out of the Yellowstone River drainage, flowing up tributaries like Tom Miner Creek. Sometime later, Member C was erupted from a vent located in the south-central part of the park near present-day Lewis Lake. These pyroclastic flows did not reach the Gardner River drainage but flowed mainly southward, toward Grand Teton National Park. Following each eruption, the region around each vent collapsed, forming one part of the overall caldera.

Although the Huckleberry Ridge Tuff was the result of the biggest of the three caldera-forming eruptions of the Yellowstone Volcano, most of the features associated with it have since been buried by subsequent eruptions. Only a short segment of the 2.1-million-year-old caldera outside the park's southwestern boundary is still preserved. In contrast, many features associated with the most recent caldera-forming eruption can still be seen today. In vignette 12, we'll take a look at the geology associated with this eruption.

12.
The Yellowstone Volcano Erupts Again!
Tuff Smothers the Region and a Caldera Forms

The Yellowstone Volcano is among the most serious potential natural hazards in North America. Fortunately, the risk that a major eruption will happen in the next few thousand years is relatively low. However, the geologic setting and history of the volcano virtually guarantee that it will experience a cataclysmic, caldera-forming eruption again. As discussed in vignette 11, the Yellowstone Volcano has undergone three caldera-forming eruptions in the past 2.1 million years. The last occurred 640,000 years ago and produced the Yellowstone Caldera. Although huge volumes of magma have poured out of the volcano since then, these younger eruptions did not form a caldera. In this vignette, we examine the Lava Creek Tuff, a rock unit that is widespread across Yellowstone and erupted immediately prior to formation of the most recent caldera. We'll also examine the best-preserved segment of the youngest caldera's margin. For more information about the origins of the Yellowstone Volcano and what drives it, see vignette 11.

To better understand how large caldera-forming eruptions develop, geologists have studied the remnants of older, presumably similar eruptions preserved along the Yellowstone hot spot track (see vignette 11). One popular idea is that a caldera-forming eruption is the climax of a longer eruptive cycle. These cycles are thought to last for about 700,000 years, the average amount of time separating the three caldera-forming eruptions of Yellowstone.

During the earliest stage of an eruptive cycle, basaltic magma begins to form about 125 miles (200 km) below Earth's surface as the uppermost mantle is heated by the thermal plume beneath Yellowstone and partially melts. The basaltic magma flows slowly upward and accumulates in a pie-shaped magma chamber at the base of the crust. Heat from the pooling magma causes the crust to expand, forming a broad topographic bulge about 200 miles (320 km) across at the surface. The granite making up the continental crust next to the magma chamber also partially melts, producing a second type of magma, called *rhyolite*. As more basalt accumulates, more rhyolite magma is generated. Eventually, an interconnected network of rhyolite magma develops in the crust, and the magma flows toward the surface. At relatively shallow depths, generally less than 10 miles (16 km) below the surface, the rhyolite magma cools to the point that most of it cannot continue flowing upward. It then accumulates in a second magma chamber that has a reservoir of molten basalt at its base. Some rhyolite magma does force its way to the surface, erupting as a series of explosive and voluminous precaldera lava flows.

During the second stage of the cycle, rhyolite magma accumulating in the near-surface magma chamber exerts pressure on the crust above

---

GETTING THERE

A good tour of the geology of the most recent caldera-forming eruption can be had between Norris and Madison Junction. To get to stop 1, drive 5.5 miles (8.9 km) south from Norris Junction on Grand Loop Road. About 0.3 mile (0.5 km) past the Beryl Spring parking lot, pull off at a small roadside pullout on the east (left) side of the road. Walk a few yards east toward the edge of the Gibbon River and look north. Be very careful of your footing at this stop, as the bank of the Gibbon River is only about 20 yards (18 m) away and is overhanging. To get to stop 2, continue south along Grand Loop Road another 1.6 miles (2.6 km). Turn right (north) into the parking lot loop, stop 2. To get to stop 3, leave the parking lot loop and proceed west on Grand Loop Road for another 0.9 mile (1.4 km) to the parking lot for Gibbon Falls. Those who would like a view across the entire caldera can proceed west on Grand Loop Road for 4.9 miles (7.9 km) to the Purple Mountain Trailhead. The hike to the summit of Purple Mountain is moderately strenuous: about 6 miles (10 km) round-trip, with a vertical climb of more than 1,500 feet (450 m). If you do set out for the summit, be sure to wear sturdy footgear and take water, extra food, and warm clothing.

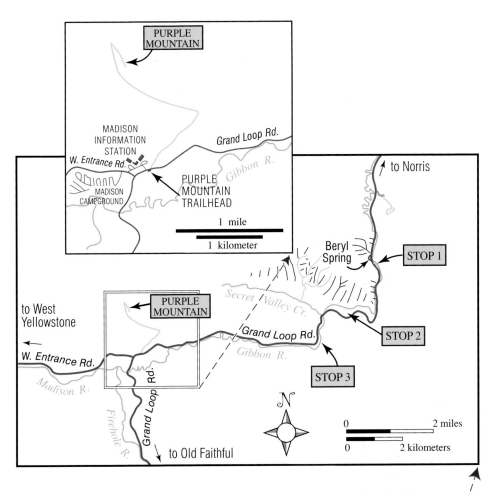

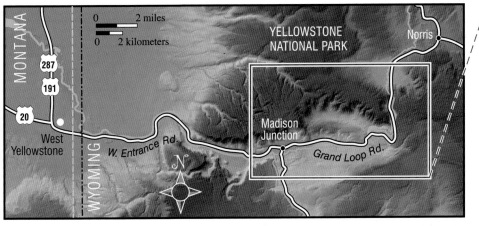

it and produces a localized bulge, typically about 10 miles (16 km) across, on the surface of the broader bulge. As more rhyolite magma accumulates, the crust around the perimeter of the localized bulge cracks, forming circular-shaped fractures called *ring faults*. The ring faults typically develop at the surface and work their way downward. Eventually, they intersect the top of the rhyolite magma chamber, allowing hot gas and ash to escape from one or more vents located along the fracture system.

The escape of gas and ash from the magma chamber leads directly to the climax of the eruptive cycle, the third stage. The escape of gas lowers the pressure within the magma chamber, allowing gas dissolved in the magma to rapidly separate from it. This process is analogous to the way a can of beer that has been shaken or is warm foams when opened. Popping the can's top releases pressure, and carbon dioxide gas that was dissolved in the beer comes out of solution, producing the foam.

The climax of the eruption is a tremendous outpouring of hot gas, ash, magma, and chunks of rock ripped off the walls of the vents located along the ring faults. The turbulent mixture is directed mostly upward

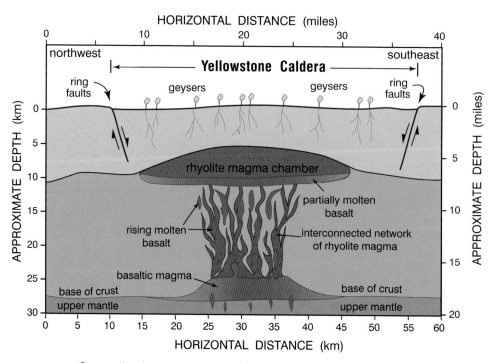

Generalized cross section of the crust under Yellowstone.

as an eruptive column that eventually collapses back to the ground. The deluge of material from the collapsing column flows outward across the landscape as a pyroclastic flow, a destructive, incandescent mixture of hot gas, ash, and magma. The ground surface within the ring faults collapses due to the volume of material expelled from the subsurface, forming a caldera.

As the eruptive cycle continues, residual rhyolite magma reaches the surface along the ring faults or other vents in the caldera floor, forming very large lava flows that partially fill the caldera (see vignette 13). The final phase of the eruptive cycle is characterized by relatively little volcanism but the widespread presence of hot springs and other thermal features. These form because groundwater is heated by the near-surface magma chamber. The slow refilling of the chamber with newly generated rhyolite magma leads to the beginning of the next cycle.

The pyroclastic flows associated with the most recent caldera-forming eruption deposited the Lava Creek Tuff across much of Yellowstone National Park. Tuff is a volcanic rock consisting of consolidated ash that either accumulates after being ejected into the air and falling back to Earth (ash-fall tuff) or is deposited by a pyroclastic flow (ash-flow tuff). The Lava Creek Tuff is composed of two members, Member A and Member B, which came from two different vents located on different parts of the Yellowstone Volcano. Stop 1 provides a good view of the top of Member A.

Look north across the road and up the canyon toward the white hot-springs deposits cropping out around Beryl Spring. Above and to the left of the white rocks and about 0.5 mile (0.8 km) away are rugged cliff-forming outcrops of orangish brown rock belonging to the upper part of Member A. In fact, the entire canyon wall across the road to the west of you consists of Member A, which in this area is more than 1,000 feet (300 m) thick.

The top part of Member A is welded tuff, a rock type that develops when a pyroclastic flow is hot enough at the time of deposition to fuse, or weld together. The pyroclastic flow that deposited this tuff was characterized by multiple surges. Surges develop due to variations in the intensity of an eruption and the volume of erupted material that falls back to Earth when the eruptive column collapses. Each surge produced one of the layers visible from stop 1.

The crude layering in Member A indicates that the pyroclastic flow that deposited it was characterized by surges; vertical cracks, or joints, in the cliffs indicate that the surges weren't separated by much time. The entire deposit probably accumulated in a matter of hours or days. The vertical cracks are called *columnar joints*, and they propagated downward in the tuff as it cooled and shrank from the top down. The fact that the joints cut through the horizontal layers in the tuff means it cooled and contracted as a single unit. That is, there wasn't enough time between each surge for a layer to cool before being buried by another layer of hot tuff from the next surge.

The columnar joints have separated the tuff into vertical columns that have been rounded by weathering. The columns are most visible in the lower part of the outcrop. If the tuff of each individual surge had cooled completely prior to arrival of the next layer of tuff, the columnar joints would not pass through the entire outcrop; instead, they would have developed only within the thickest individual layers, if at all.

Once you've had a chance to see the exposures of Lava Creek Tuff from stop 1, proceed south along Grand Loop Road to stop 2. From

Member A of the Lava Creek Tuff with crude columnar joints, as seen from stop 1. The horizontal layers reflect individual surges of the pyroclastic flow that deposited the tuff.

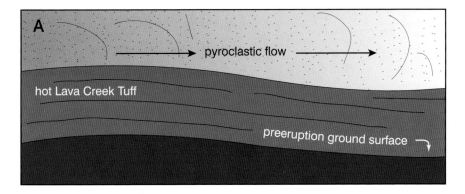

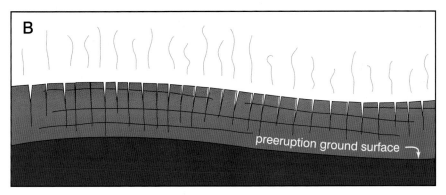

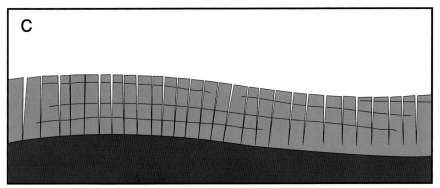

How the columnar joints at stop 1 formed. (A) Layers of hot tuff accumulate from pyroclastic flows and begin to cool. (B) Vertical cracks propagate downward as the tuff cools from the top down. (C) Eventually, the intersecting vertical cracks extend through the entire unit and break it into a series of vertical columns. The columns typically have six sides but can have as few as three or as many as twelve. The number of sides is related to how quickly the tuff cools, the physical strength of the rock, and the shape of the deposit. The columns are subsequently accentuated and rounded by weathering.

stop 2, look north back toward the prominent outcrop of rock on the far side of the Gibbon River. This is also Member A. Unlike the outcrop at stop 1, this exposure of the Lava Creek Tuff does not display good columnar jointing because numerous fractures in the rock face cut across the joints, obscuring them. The fractures exist because the rock face is the surface expression of a major fault. You are standing on a big block of Lava Creek Tuff that dropped downward along the fault relative to the rock face. The fault runs parallel to the face and dives into the subsurface at a steep angle.

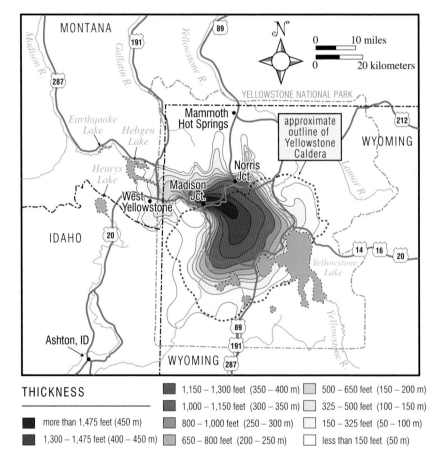

THICKNESS

more than 1,475 feet (450 m)
1,300 – 1,475 feet (400 – 450 m)
1,150 – 1,300 feet (350 – 400 m)
1,000 – 1,150 feet (300 – 350 m)
800 – 1,000 feet (250 – 300 m)
650 – 800 feet (200 – 250 m)
500 – 650 feet (150 – 200 m)
325 – 500 feet (100 – 150 m)
150 – 325 feet (50 – 100 m)
less than 150 feet (50 m)

Distribution of Member A of the Lava Creek Tuff in Yellowstone Country. The tuff is thickest near Purple Mountain, north of Madison Junction, and tapers to the east. These relationships suggest that the vent from which Member A erupted was probably near Purple Mountain and that most of the pyroclastic flows were directed to the east.

This fault formed shortly after the Yellowstone Volcano collapsed, but it is not the fault that defined the original northern edge of the caldera. The original caldera wall was about 1 mile (1.6 km) south of where you are standing. The thick, still-hot Lava Creek Tuff was not strong enough to maintain the steepness of the original caldera wall, so a giant slice of tuff slid down along the fault in front of you, exposing the steep rock face shortly after the caldera formed. Stop 2 is located on this giant mass of displaced rock, called a *slump block*, which is more than 1 mile (1.6 km) across and about 4 miles (6.4 km) long.

From stop 2, proceed west on Grand Loop Road to Gibbon Falls (stop 3). Between stops 2 and 3, you are traveling on the slump block. At stop 3 rounded columns of welded tuff are clearly visible across the river. As with those seen at stop 1, these are defined by columnar joints that developed in Member A of the Lava Creek Tuff. Gibbon Falls exists where the river flows off the south side of the slump block and into the caldera. The river has cut into the slump block, carving out the canyon containing Gibbon Falls and causing the waterfall to migrate 0.25 mile (0.4 km) upstream from its original position, which was also the original position of the slumped crater wall. The surface down which the water flows is an erosionally modified remnant of the original crater wall.

Looking north from stop 2. This steep, strongly fractured rock face is made of Member A of the Lava Creek Tuff and was exposed when a large segment of the newly formed caldera wall slid into the crater, forming a slump block. Stop 2 is located on the slump block.

The original position of the slumped crater wall projected from east to west across the mouth of the Gibbon River canyon downstream. Most of the caldera has since been filled in by younger rhyolite lava flows, including the Nez Perce Creek flow that makes up the rolling hills beyond the mouth of the canyon to the south, so you have to use your imagination to visualize the presence of the caldera's edge from

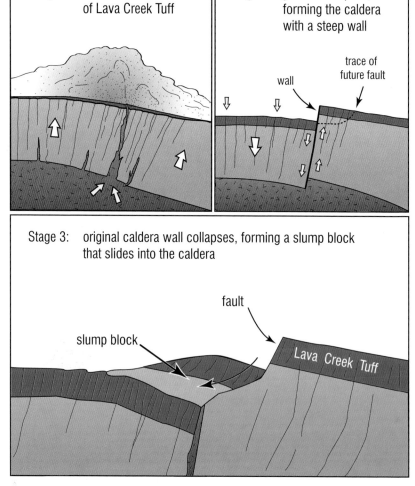

Following the eruption of the Lava Creek Tuff and formation of the Yellowstone Caldera, a large segment of the steep north face of the caldera slid into the crater, forming a slump block. Stop 1 is located north of the caldera margin, and stops 2 and 3 are located on the giant slump block.

here. It helps to mentally remove the low hills corresponding to the Nez Perce Creek flow as you look southward.

In vignette 13, we'll turn our attention to the enormous rhyolite flows that erupted after the last climactic, caldera-forming eruption. These lavas, representative of Yellowstone's most recent volcanic activity, were so voluminous that they mostly filled in the caldera and reached well beyond its margins in places. The flows cover much of the heart of Yellowstone and are a key part of Yellowstone's volcanic history.

Gibbon Falls occurs where the river cascades down the eroded, tilted southern side of a slump block and into the caldera crater. Note the columnar joints on the far (eastern) side of the river.

Purple Mountain Hike

The trail to the summit of Purple Mountain climbs up the eroded northern margin of the Yellowstone Caldera, and the summit offers a great view to the south across the caldera. Few outcrops occur along the trail, which first climbs up a wooded hillside that formed on the Mount Jackson Rhyolite. This rhyolite erupted prior to the Lava Creek Tuff. Higher up, the trail crosses a contact between the rhyolite and the lower-most part of the Lava Creek Tuff's Member A. Unlike the densely welded tuff that forms the tall cliffs and prominent rounded columns of the top part of Member A, seen at stop 1, the lower part of Member A generally does not crop out well because it was not welded, or fused together, by residual heat after it was deposited. As a result, the tuff here is a lower-density rock that is more easily broken down by the elements. The upper part of Member A does not occur on Purple Mountain.

At the summit there is a good view of the slump block that collapsed into the caldera. The slump block is the low timbered ridge about 2 miles (3.2 km) to the east. It is located between the Gibbon River to the south and the prominent escarpment to the north, which is the fault down which the slump block slid into the caldera crater. The caldera margin runs westward along the south side of the slump block, passes along the foot of Purple Mountain, and continues to the west and southwest, crossing the Madison River before curving to the south.

The low-elevation country on the far (south) side of the Gibbon River dropped when the caldera formed. The original caldera crater has since been almost entirely filled in by younger rhyolite flows, and these flows form the low, rolling hills across which you are looking. On a clear day, you might spot plumes of steam rising from Lower Geyser Basin, located along the Firehole River on the caldera floor, or parts of the Teton Range, including Grand Teton, located on the far side of the caldera about 25 miles (40 km) to the south.

Looking west along Grand Loop Road between Gibbon Falls and the Purple Mountain Trailhead. Though eroded back a bit from their original location, the steep, south-facing slopes on the skyline mark the northern edge of the caldera, which the road parallels. The pinkish rock cropping out next to the road is Lava Creek Tuff.

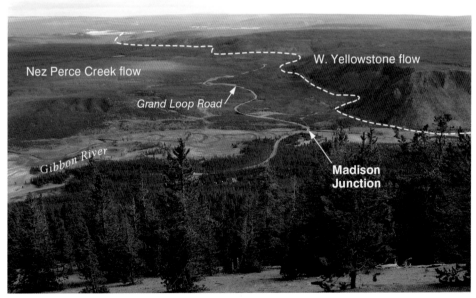

Looking south from the summit of Purple Mountain across the central part of the Yellowstone Caldera. The dashed white line represents the boundary between the Nez Perce Creek rhyolite flow, which erupted about 160,000 years ago, and the West Yellowstone rhyolite flow, which erupted about 108,000 years ago. We'll visit this location in the next vignette.

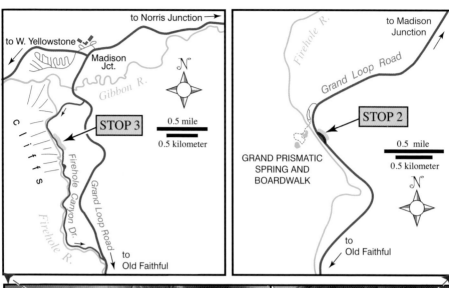

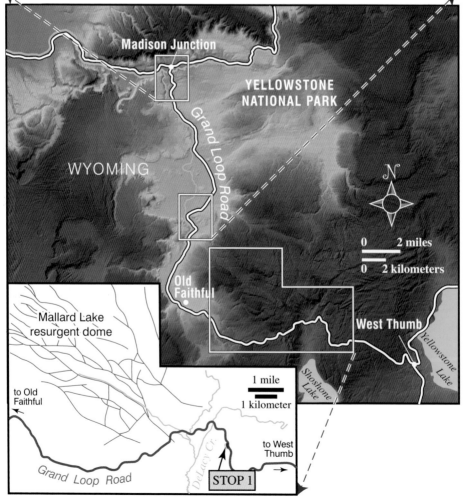

13.
The Youngest Eruptions
Rhyolite Flows in the Firehole River Drainage

Since the eruption of the Lava Creek Tuff and formation of Yellowstone's most recent caldera 640,000 years ago, the Yellowstone Volcano has continued to be very active. Tremendous volumes of rhyolite magma, along with lesser amounts of basaltic magma, have erupted from vents located on various parts of the volcano. These magmas cooled to form the volcanic rocks rhyolite and basalt, with rhyolite being richer in

---

GETTING THERE

Stop 1 offers a spectacular view of the Mallard Lake resurgent dome. From West Thumb Junction, proceed northwest on Grand Loop Road for 6.8 miles (10.9 km) to a small roadside pullout on the west (left) side of the road. If you are coming from the other direction, the pullout is 10.8 miles (17.4 km) past the overpass leading to Old Faithful. The small hill overlooking the pullout provides the best views of the dome, which is to the northwest. Stop 2 offers a cross-sectional view of the Biscuit Basin rhyolite flow. To get there, follow Grand Loop Road northwest from stop 1 for 16.3 miles (26.2 km) to the pullout on the east (right) side of the road; it is 0.3 mile (0.5 km) south of the main parking area at Lower Geyser Basin. The outcrop you want is at the north end of the pullout. To reach stop 3, drive 9.8 miles (15.8 km) north along Grand Loop Road and turn left (west) onto the one-way Firehole Canyon Drive. Proceed 0.8 mile (1.3 km) to the small pullout on the left (east) side of the road just beyond a prominent roadcut on the same side. Walk along the road uphill and to the south for 500 feet (150 m) or so to examine the front of the Nez Perce Creek flow.

---

silica than basalt. The post-caldera rhyolite flows were so voluminous that, collectively, they filled in much of the caldera, and its original rim is exposed in only a few places, most notably between Gibbon Falls and Purple Mountain (discussed in vignette 12).

Geologists with the U.S. Geological Survey have mapped forty individual post-caldera rhyolites across Yellowstone. These range in size and shape from relatively small domes less than 1 square mile (2.6 square km) in area to gigantic flows exceeding 100 square miles (260 square km). Most are so large that it is difficult to appreciate their overall shape from the ground. The biggest individual flows are about 20 miles (32 km) across and nearly 1,000 feet (300 m) thick and cover large portions of the caldera floor. Others spilled out of the caldera and down the western flank of the volcano, reaching up to 10 miles (16 km) beyond the boundary of Yellowstone National Park.

The rhyolite flows constitute part of the long-term eruptive cycle of the Yellowstone Volcano (discussed in vignette 12) and have greatly shaped many of the landscapes within Yellowstone Country. Not only is the relatively subdued topography within the heart of the park a direct result of the flows, but weathering of the rhyolite has produced thin, nutrient-poor soil unable to support diverse tree types. As a result, Yellowstone's forests are almost exclusively lodgepole pine and are very susceptible to large wildfires—a topic we'll examine in the epilogue. In this vignette, we'll look at an area of the caldera floor that has been deformed by magma working its way upward, and then we'll turn our attention to some of the rhyolite flows, examining their structure and considering what their geology suggests about past environmental conditions.

Geologists have determined the approximate locations of the vents from which the rhyolite flows erupted by mapping the shape of each flow and the orientation of flow lines. Flow lines are broad folds that develop on top of lava flows perpendicular to the direction the lava moves. They form as the upper surface of the flow congeals while less viscous molten lava below continues flowing, deforming the upper surface.

The vents are clustered on different parts of the Yellowstone Volcano. One cluster occurs along the southwestern and western segments of the caldera's ring faults, about 7 miles (11.3 km) southwest of Old

Faithful. A second cluster is located within the caldera in the Mary Mountain region, roughly in the center of the loop formed by Grand Loop Road. Additional vents are located within the caldera floor on or very close to two elliptically shaped resurgent domes. Resurgent domes typically form in the floors of calderas or along ring faults following a caldera-forming eruption. Magma congeals over and around vents or fractures, and pressurized magma below causes the ground to bulge upward. Each of Yellowstone's resurgent domes is roughly 10 miles (16 km) long, 7 miles (11.3 km) across, and 1,000 feet (300 m) above the surrounding landscape. One is located less than 4 miles (6.4 km) north of Fishing Bridge and is called the Sour Creek Dome. The other, called the Mallard Lake Dome, is the same distance east of Old Faithful and is well displayed to the northwest from stop 1.

Geophysical images have shown that there is a large magma chamber in the crust beneath the Yellowstone region. Geophysicists produced the images by analyzing data from an extensive network of seismographs deployed across the area. These machines detect and measure earthquake waves, which travel through the Earth away from the site of a temblor. When the waves pass through partially molten rock, they slow down. The magma chamber under Yellowstone is not an underground reservoir of liquid magma surrounded by solid rock. Rather, it is a region of solid rock with an interconnected network of liquid magma. Only 1 to 2 percent of the rock in the magma chamber is molten; the rest is hot, but solid. The top of the chamber has two prominent vertical bulges, each located beneath one of the two resurgent domes.

Using a sophisticated technique called *laser interferometry*, scientists with the U.S. Geological Survey have shown that both resurgent domes, as well as the caldera floor between them, have experienced substantial vertical ground movement over the past twenty years. For this type of analysis, a satellite passing overhead measures the elevation of the ground surface by bouncing radar waves off it. The same measurements are made a year or more later, and precise changes in the ground surface elevation can be determined by comparing the two sets of data. Between 1992 and 1993, the Sour Creek Dome subsided a little more than 1 inch (2.5 cm), and by 1993 the Mallard Lake Dome began subsiding—about 1.5 inches (3.8 cm) between 1993 and 1995. Between 1995 and 1996, the Sour Creek Dome reversed direction and rose about 1 inch (2.5 cm), while lesser amounts of uplift occurred on the caldera floor between the two domes. At the same time, subsidence of the Mallard Lake Dome slowed and nearly stopped.

This relatively recent ground motion in specific parts of the caldera floor has occurred against a backdrop of vertical movement across larger parts of the caldera floor and regions outside the caldera. By comparing results from two leveling surveys, scientists discovered that much of the caldera floor rose more than 3 feet (90 cm) between 1923 and 1975. Additional leveling surveys showed that this uplift ceased abruptly in 1984, and for a period of about two years the caldera floor stopped moving. Then it began to sink at about the same rate it had been rising, dropping about 7 inches (18 cm) in ten years.

The vertical ground motions are driven by the movement of magma and gas in the magma chamber. One region may subside as magma and gas move away from it, while the area the magma moves to will rise. The two prominent bulges in the top of the magma chamber suggest

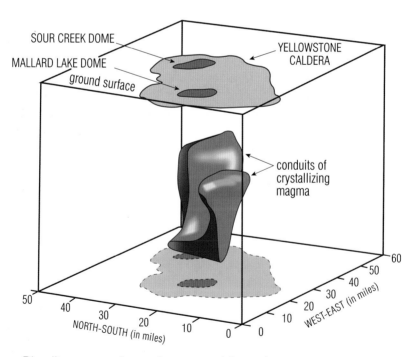

Rhyolite magma is moving upward from the magma chamber under the park via two large conduits located below the Sour Creek and Mallard Lake resurgent domes. Although this image appears to represent a magma chamber full of liquid, only about 1 to 2 percent of the rock is molten.

that a great deal of magma and gas is working its way upward through a network of cracks and fissures between the magma chamber and the resurgent domes above. Multiple large rhyolite flows have erupted from the resurgent domes, suggesting that the bulges are connected to conduits through which the magma passed on its way to the surface. Geologists have suggested that the conduits may have been in existence since the caldera formed, if not before.

Both resurgent domes are crisscrossed by normal faults. These formed due to extension in the crust as it bulged upward. The upward bowing caused blocks of crust to drop downward along the axes of the domes, much like a keystone block in a Roman arch would settle downward if the two sides of the arch were pulled apart. Such a down-dropped fault block is called a *graben*, which is German for "ditch." Although some grabens are bounded on either side by single faults, more commonly multiple parallel faults occur on either side of a graben.

The view to the northwest from stop 1 looks directly down the axis of a large graben that formed on the Mallard Lake Dome. You may want to climb up the small hill above the pullout for the best view. Take a look at the hills forming the skyline to the northwest on the far side of the DeLacy Creek valley. The prominent, northeast-facing slope, which is tilted to the right, is the northeastern side of the Mallard Lake Dome. The lower-elevation hills to the left of this slope mark the graben. These hills comprise the top of the fault block (keystone) that dropped downward along faults on either side of it. The southwestern flank of the dome is defined by the gentle slopes located above prominent roadcuts. The short, steep, east-facing slopes (tilted to the right) above the roadcuts define normal faults between the southwestern flank of the dome and the central graben.

Looking northwest from stop 1 at the Mallard Lake Dome and the graben that formed along its axis.

At least three pulses of rhyolite magma have come from vents located on or very close to the Mallard Lake Dome. The earliest of these erupted 516,000 years ago from a vent (maybe multiple vents) located within 1 mile (1.6 km) of the northwestern end of the dome. Called the Biscuit Basin flow, it was a relatively small flow—or series of flows—with a combined volume of about 1 cubic mile (4 cubic km). Following were the Scaup Lake flow, which erupted 198,000 years ago, and the Mallard Lake flow, which erupted 151,000 years ago. The rocks forming the Mallard Lake Dome belong almost entirely to these two younger flows.

Geologists have grouped all of the post-caldera rhyolite flows strati-graphically into the Plateau Rhyolite, which in turn is subdivided into members, each made of at least one mappable flow. The oldest member is the Upper Basin Member, which includes both the Scaup Lake and Biscuit Basin flows. The Biscuit Basin flow is the first known rhyolite flow to have erupted following the climactic eruption of Lava Creek Tuff and the formation of the caldera. Many geoscientists view the Upper Basin Member as representing the final depletion of magma related to the climax eruption of the Lava Creek Tuff.

The Biscuit Basin flow at stop 2. Hammer *(circled)* for scale. The bottom of the outcrop is well layered but the layering is much less distinct toward the top, which is fragmented.

An excellent outcrop of the Biscuit Basin flow can be examined at stop 2. From the pullout, walk north along the shoulder of the road to the gray and orange outcrop, which is about 30 feet (9 m) high. It is best to first view the outcrop in its entirety from a position on the far (west) side of the road. The outcrop presents a cross-sectional view of the flow; only the very base of the flow is not exposed. The dark gray rhyolite at the base of the outcrop is clearly layered, but the orangish rhyolite forming the top is more fragmented and lacks well-developed layering. These characteristics reflect differences in the nature and behavior of the rhyolite as it flowed slowly across the landscape from a vent less than 1 mile (1.6 km) to the northeast. Cooling in the upper part of the flow caused a crust to form, and continued movement of the flow beneath caused the crust to fragment. The relatively hotter interior did not fragment; rather, it deformed like taffy.

The obvious dark- and light-colored streaks at the base of the outcrop are called *flow bands*. Although geologists don't fully understand how flow bands develop, some studies have concluded that they form due to variations in the composition or abundance of crystals or bubbles within the lava. Incomplete mixing within the flow causes these differences to be smeared out as bands. The thick consistency of the lava preserves the flow bands, much like the swirls that would remain if a cup of melted chocolate was incompletely stirred into cake batter.

In several places the flow banding defines impressive tight folds in the rock. These folds represent internal deformation the viscous lava experienced as it flowed across the land surface. Friction caused by the irregular ground surface caused the flow to fold over on itself. The most visible folds are about 1 foot (30 cm) across and occur just above the base of the outcrop, about halfway between its northern and southern ends. Larger folds more than 5 feet (1.5 m) or so across are present a little higher in the outcrop, but these are less obvious. The folds indicate that the lower parts of the flow behaved in a ductile fashion and did not fragment like the upper parts of the flow.

In addition, the Biscuit Basin flow at stop 2 contains an abundance of marble- to ping-pong-ball-sized spherical features called *lithophysae*, Greek for "rock bubble." The formation of lithophysae is not well understood, but some geologists have suggested that they form due

Tight folds like these are common near the base of the Biscuit Basin flow at stop 2. The folds range in size from the small ones shown here to more than 5 feet (1.5 m) across. Lens cap for scale.

to the alteration of volcanic glass in the flow. The alteration starts at a single point and spreads outward, forming a small spherical object: a lithophysa. Within the zone of alteration—that is, within each lithophysa—the molecules forming the glass rearrange into microscopic mineral crystals with tiny pore spaces. As the alteration spreads outward, the lithophysa grows.

Other geologists have suggested that the lithophysae represent bubbles that formed in the cooling magma. Finely crystalline feldspar, quartz, and other minerals precipitate as concentric shells from the vapors in the gas bubble. Most of the lithophysae at stop 2 have a mineral lining that may have formed in this manner. The lithophysae are absent in the dark gray rhyolite at the base of the outcrop on its southern end but increase in abundance upward within the flow. Bubbles are usually more abundant near the top of a lava flow because less pressure exists there than at deeper levels. Anyone who has dived to the bottom of a deep swimming pool has experienced a similar increase in pressure with depth.

The dark gray rhyolite in the lowest part of the outcrop contains an abundance of clear or white crystals of the feldspar mineral oligoclase. Rather than having formed in the cooling lava flow, these crystals were probably part of the magma when it erupted. Studies of similar crystals in other flows of the Upper Basin Member have shown that the ages

Lithophysae in the rhyolite at stop 2. Pen tip for scale.

of individual crystals vary widely despite their similar appearance and the fact that they occur in the same rock. The variety of ages indicates that most were introduced to the magma from other sources, probably older rhyolites forming the vent walls that the magma passed by during the eruption. Such crystals—those that did not form in the magma they were erupted in—are called *xenocrysts*, from the Greek *xeno*, meaning "stranger."

The lower portion of the outcrop also contains an abundance of BB-sized granules of black obsidian, a type of volcanic glass. Unlike minerals, volcanic glass lacks an orderly molecular structure. It forms when magma cools too quickly for minerals to develop. Geologists from the U.S. Geological Survey have described this part of the Biscuit Basin flow as perlite, a type of volcanic glass containing high amounts of water and characterized by small concentric fractures that cause the rock to resemble a mass of small spheroids. The spheroids are thought to form when the rhyolite lava contracts due to very rapid cooling, or quenching. In fact, the presence of the perlite has led the U.S. Geological Survey geologists to speculate that the Biscuit Basin flow may have run into a lake in the caldera floor, which quenched the lava and provided the high water content of the glass.

Following the eruption of the Biscuit Basin flow and other flows of the Upper Basin Member, the Yellowstone Volcano underwent a period of relative inactivity beginning around 486,000 years ago and lasting for about 321,000 years. Relatively few eruptions happened, although the Scaup Lake flow is a notable exception. Then, around 165,000 years ago, the volcano woke up, beginning a period of massive rhyolite eruptions. These occurred in three distinct pulses, ending 70,000 years ago—the age of the youngest known eruption. Most of these younger rhyolites belong to the Central Plateau Member. The rhyolite flows of this member have distinctly different compositions than those of the Upper Basin Member, and some geologists regard them as representing a new eruptive cycle of the Yellowstone Volcano.

The Central Plateau Member includes several truly gigantic flows, many of which spread well beyond the caldera's margins. Among these are the Nez Perce Creek flow and the West Yellowstone flow, both of which can be viewed at stop 3. The Nez Perce Creek flow erupted about 160,000 years ago from a vent in the Mary Mountain region. It traveled mostly west, and the front of the flow reached just beyond the Firehole River, more than 12 miles (19 km) from the vent. The

The northeastern front of the West Yellowstone flow as seen from stop 3. The physical mixing of lava in the flow created the chaotic appearance of the rock on the right side of the photo.

West Yellowstone flow erupted 108,000 years ago from a vent near the southwestern edge of the caldera, very close to Plateau Lake. This small lake in the remote west-central part of Yellowstone National Park occurs near an 8,855-foot (2,699 m) hill that formed as the last bits of magma left the vent and solidified. From its vent, the West Yellowstone flow ran about 15 miles (24 km) to the northeast, reaching nearly all the way to Madison Junction.

The western end of the Nez Perce Creek flow impeded the progress of the northeastern end of the West Yellowstone flow. From stop 3, the tall canyon wall across the Firehole River is the front of the West Yellowstone flow. Parts of the wall appear chaotic, reflecting the mixing of the moving lava along the flow front.

At stop 3 you are actually standing on the front of the Nez Perce Creek flow, and the rocks exposed along the east (left) side of the road provide an excellent glimpse of the flow's internal structure. Contorted, nearly vertical layers of rhyolite reflect the ductile movement of the hot, viscous lava as it was smeared out within the flow. Although these

The internal structure of the Nez Perce Creek flow at stop 3. The layers that slope from the upper left to lower right formed as lava was smeared out within the flow. Much of the rest of the lava is a rhyolite breccia.

layers are striking, most of the rock forming the outcrop is a rhyolite breccia made of angular clasts ranging from baseball-sized to the size of a microwave oven. The angular clasts developed as the cooled outer portions of the flow fragmented and were reincorporated into the still-moving lava. Good views of the breccia can be seen on the northern end of the roadcut.

A second rhyolite breccia that reflects the highly explosive nature of the eruptions of the Nez Perce Creek flow is exposed at the southern end of the roadcut, about 85 yards (75 m) up the canyon from the pullout. Here, angular clasts of black obsidian are encased in a light-colored, ash-rich matrix. The fact that the clasts are made of obsidian suggests that the lava cooled very rapidly, and the angularity of the clasts suggests that this process was violent enough to fragment the quenched lava.

Some of the rock on the southern end of the outcrop at stop 3 is perlite—obsidian preserved as masses of BB-sized spherules. The presence of perlite and the fact that it occurs in breccia has led geologists with the U.S. Geological Survey to deduce that the Nez Perce Creek flow encountered a body of cold water, or possibly glacial ice, within the caldera. The possibility conjures images of an impressive scene in

The angularity of these clasts within an ash-rich matrix at stop 3 suggests that the lava flow was at times explosively fragmented. The clasts are black because they are made of obsidian, which formed as the hot rhyolite lava was quenched. Rock hammer for scale.

which hot lava—estimated to be around 1,560 degrees Fahrenheit (850 degrees C) when it erupted—was rapidly quenched to form perlite and also underwent a series of localized explosions when it reached the water or ice, forming huge clouds of hissing steam and creating the angular blocks of black obsidian.

Supporting evidence for violent interactions between lava and ice in the caldera is found along the eastern margin of the West Yellowstone flow, which has a sheath of perlite along its length. The eastern margin also contains several odd reentrants, or indentations, that are not typical of rhyolite flows and are not characteristic of the western margin of the West Yellowstone flow itself, which is marked by more typical tongue-shaped lobes. The most prominent reentrants occur near Little Firehole Meadows, Buffalo Meadows, and Lower Geyser Basin. Both the reentrants and the perlite sheath could have formed if the lava encountered lobes of glacial ice. Recently published analyses of the

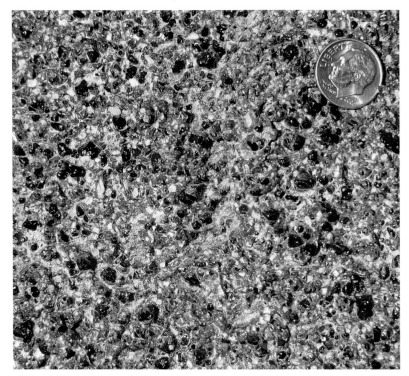

Close-up of perlite in the Nez Perce Creek flow at stop 3. The presence of perlite, and the fact that it commonly occurs as angular clasts in the flow, has led some geologists to infer that the lava encountered a body of water, or possibly glacial ice, in the caldera. Dime for scale.

effects of glaciation in the Yellowstone region suggest that large glaciers had formed and reached their maximum size as recently as 11,000 years before the West Yellowstone flow erupted. Perhaps remnants of these glaciers still existed.

Although no one knows with certainty whether glacial ice caused the perlite and large reentrants to form on the front of the West Yellowstone flow, glacial ice has profoundly shaped many of the modern landscapes in Yellowstone Country. In vignettes 14, 15, and 16, we turn our attention to the relatively recent history of glaciation in the region.

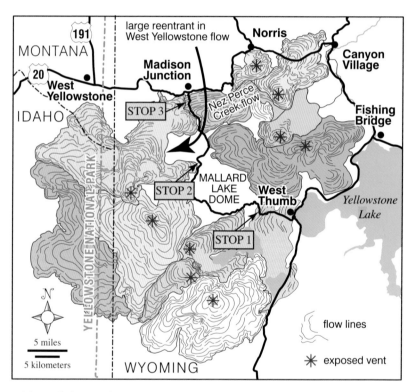

Map of the major post-caldera rhyolite flows and their vents. The eastern margin of the West Yellowstone flow has several major reentrants, the biggest labeled here. Geologists have interpreted these reentrants, along with the presence of a sheath of perlite along the flow front, as suggesting that the rhyolite flow encountered glacial ice in the caldera.

14.

Ice Sculptures along the Beartooth Highway

Glaciers Carve Yellowstone's Landscapes

The Yellowstone region is world famous for its enchanting landscapes. Sharp mountain peaks rise up next to U-shaped valleys; randomly strewn, humongous boulders litter rolling hills known for herds of bison and elk; wide, grassy meadows containing meandering streams invite fly-fishers of all levels. Each of these scenes has been partially shaped by glaciers that grew, advanced, and then melted back across the Yellowstone region more than once during the past 200,000 years. In this vignette and vignettes 15 and 16, we'll turn our attention to the history of glaciation in Yellowstone and the effects that glaciers had on its landscapes. This vignette focuses on erosional features glaciers formed as they scraped and slid their way across Yellowstone Country.

Multiple times during the past 2 million years, global temperatures dropped low enough for a long-enough period of time that large glaciers formed on many of the continents, particularly in regions of high latitude and high elevation. Most of these cold periods probably developed due to seasonal changes in the amount of sunlight reaching different latitudes. These changes were brought on by variations in the shape of Earth's orbit around the sun, changes in the tilt of Earth's axis, and a periodic wobbling of Earth's axis that is similar to the way a top wobbles before it stops spinning. When the amount of sunlight changed such that summers became relatively brief and cool, glaciers formed.

How do glaciers form? Shortly after snow falls and accumulates in a snowpack, it begins to change in response to daytime warming and nighttime cooling. The delicate six-sided snowflakes seen in freshly fallen snow become rounded as their tips melt and then refreeze. These changes, combined with the compaction of the snowpack, cause the snow to become denser. Eventually, it transforms into rounded crystals of ice called *firn*. Small granules of firn can be seen in late spring and early summer in snowfields that remain along some of the higher-elevation roadways in and out of the park. The following winter, the firn is buried deeper in the snowpack by more snow. Additional compaction, partial melting and refreezing of the firn, and freezing of water that percolates into the snowpack form glacial ice.

Once glacial ice accumulates to a great-enough thickness, it can move downslope. Scientists have shown that glaciers flow due to internal deformation. Much as water in the middle of a river flows faster than that along the banks, where the friction is greater, ice in the middle of a glacier flows faster than ice at its margins. In studies in which a series of poles were driven into glacial ice to form a straight

GETTING THERE

Spectacular examples of the power of glacial ice as an erosional agent are plainly visible along the Beartooth Highway—one of North America's most scenic stretches of road. To reach stop 1 from Red Lodge, Montana, start at the intersection of US 212 and Ski Run Road, located at the south end of town. Proceed 28.2 miles (45.4 km) southwest on US 212 to a switchback (one of many) that curves sharply to the left. It is the first switchback past the parking lot to the Gardner Lake Trailhead, part of the Beartooth Loop National Recreation Trail. If you are coming from Yellowstone National Park or Cody, Wyoming, stop 1 is 20.9 miles (33.6 km) northeast of the intersection of US 212 and Wyoming 296, or 1.4 miles (2.3 km) past the marked west summit of Beartooth Pass. Once you have parked, carefully cross the road and step over the guardrail. Stop 2 is the boat launch for Beartooth Lake. From stop 1, proceed 11.8 miles (19 km) southwest on US 212 (9.1 miles, or 14.6 km, northeast of the US 212 and Wyoming 296 intersection), turn right into the Beartooth Butte Campground and continue about 0.3 mile (0.5 km) to the boat launch. Stop 3, on US 212, is 2.3 miles (3.7 km) southwest of the turnoff to the Beartooth Butte Campground, or 6.8 miles (10.9 km) northeast of the intersection of US 212 and Wyoming 296. Stop 4, a roadcut on the north side of Wyoming 296, is 5.7 miles (9.2 km) southeast of its intersection with US 212.

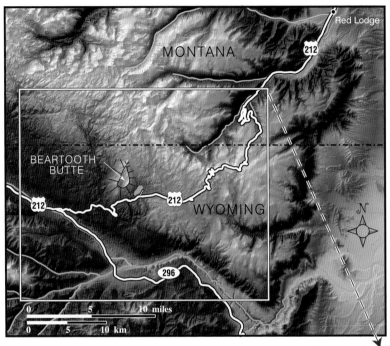

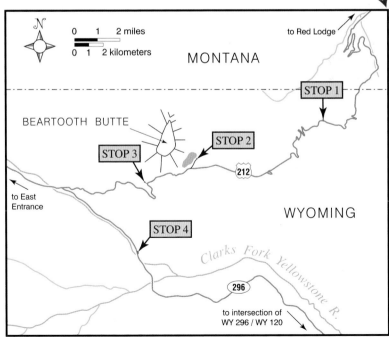

line stretching across a glacier, the poles in the center of the glacier were carried downslope faster than the poles on the edges. However, all of the poles moved downslope relative to a set of reference poles that were driven into the ground on the sides of the glacier.

In addition to flowing like a very thick fluid, glaciers also slide downslope, like a block of ice on a tilted board. Sliding occurs more easily when water accumulates beneath a glacier and lubricates its base, in the same way that a block of wet ice will slide down a tilted board faster than a block of ice that is dry. Combined, sliding and internal flowing cause glaciers to move at rates ranging from less than 1 inch (2.5 cm) per day to more than 1 foot (30 cm) per day, although occasionally glaciers will surge more than 100 yards (90 m) in a day.

Although the term Ice Age is commonly used to describe times of widespread glacial growth, there is no such formally recognized interval of geologic time; rather, periods in which glaciers advanced enough to leave behind a geologic record are called *glaciations*. Glaciers became widespread across Yellowstone at least twice since the Lava Creek Tuff was erupted 640,000 years ago. The older of the two glaciations is called the Bull Lake glaciation, named after a lake in the Wind River Range of Wyoming where evidence of this glacial advance is well preserved. The younger episode is called the Pinedale glaciation, named after a small town on the western flank of the Wind River Range.

Although geologists have long recognized that sediment deposits related to the Bull Lake and Pinedale glaciations are widespread across much of the northern Rocky Mountains, considerable controversy and discussion have surrounded the timing of the two glaciations, particularly that of the Bull Lake glaciation. Recently, this controversy has largely been eliminated. Using a technique called *cosmogenic surface exposure dating*, geoscientists have determined when both glaciations occurred in the Yellowstone region.

To date rock using this method, geoscientists remove a small sample from the upper surface of a glacially deposited boulder and then isolate pieces of quartz from the sample in the lab. Using sophisticated instrumentation, they then measure the amount of beryllium-10 in the quartz. Beryllium-10 is a substance that forms and accumulates in quartz when cosmic rays from outer space strike the mineral. By determining the amount of beryllium-10 that has accumulated, scientists are able to calculate how long the boulder has been exposed—that is, the amount of time the boulder has been receiving cosmic radiation since being deposited by a glacier. In the Yellowstone region, Bull Lake deposits contain boulders ranging between 157,000 and 123,000 years old, averaging about 136,000 years. Boulders sampled from Pinedale glacial deposits range between 18,000 and 16,000 years old. (We'll examine the deposits of glaciers in more detail in vignette 15.)

Snow and ice first began to accumulate and form valley glaciers in the higher mountainous regions of Yellowstone, including the Beartooth, Absaroka, Gallatin, and Washburn ranges. As these glaciers grew, they flowed outward from each range. During the Pinedale glaciation, the glaciers grew to be so extensive and so thick that they formed a connected mass of ice that stretched completely across the Yellowstone Plateau—the high-elevation region in Yellowstone that contains the subdued topography of the caldera and flanks of the surrounding volcano. This ice merged with a similarly large mass of ice that formed over the Beartooth Plateau—the high, relatively flat topography found in the heart of the Beartooth Mountains. Collectively, these ice masses formed the Yellowstone Ice Cap. At its maximum, the ice cap stretched uninterrupted from the Teton Range south of Yellowstone National Park across the entire park and over the Beartooth Plateau to the north, covering everything but the highest mountain peaks. In all of Yellowstone National Park, only a small area along the park's western boundary was not covered by the ice cap.

It probably took at least a few thousand years, and possibly 10,000 years, for the ice cap to form. This estimate is based on assumptions about the amount of precipitation that fell each year, combined with evidence regarding the thickness of the Yellowstone Ice Cap determined from the mapping of glacial features in the area. For example, it would take at least 1,000 years for 2,000 feet (610 m) of ice to form if a hypothetical 2 feet (60 cm) of precipitation, expressed as water, fell each year and all of it became glacial ice. Because not all of this precipitation would have become glacial ice, and because glacial ice flows laterally and does not simply build up vertically, the 1,000-year estimate is a minimum value.

Ice within the Yellowstone Ice Cap flowed outward, away from the crest of its surface toward its margins. It generally flowed down preexisting river valleys, but in some areas the ice was thick enough to completely overwhelm smaller hills and mountains, flowing up and over them. As it flowed along, the ice dislodged large pieces of fractured bedrock, picking them up and carrying them downstream.

The many fragments of rock carried by glacial ice, combined with the weight of the ice, act like a giant piece of sandpaper bearing down on the bedrock below. Some geoscientists use the term "glacial buzz saw" to describe the tremendous power glaciers have to erode bedrock in mountainous terrain. In fact, there is an ongoing scientific debate

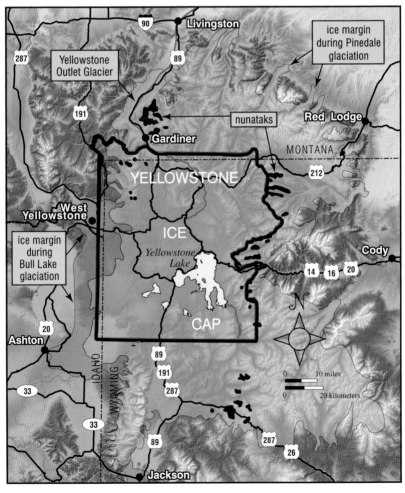

The Yellowstone Ice Cap about 17,000 years ago, during the Pinedale glaciation. The nunataks were peaks that stuck out above the ice cap.

regarding the glacial buzz saw theory—the idea that glaciers and their erosive power limit the overall elevations to which mountain ranges would otherwise grow by tectonic forces.

One of the best places to view the powerful erosional effects of glaciers is at stop 1, a sharp switchback along US 212 near Beartooth Pass. From Red Lodge, US 212 gains more than 4,700 feet (1,430 m) of elevation and passes from the former margin of the Yellowstone Ice Cap (near Red Lodge) to the heart of the Beartooth Plateau, where stop 1 is located. The rugged mountains and deep canyons laid out before you at this stop create one of the most impressive views in all of Yellowstone Country. Virtually all of the valleys and ridges you can see are dark gray and brown Precambrian igneous and metamorphic basement rock of the Beartooth Plateau. This crystalline gneiss and granite is generally between 3.1 and 2.8 billion years old (see vignette 1). The rocks are at high elevations today because of uplift that occurred about 60 million years ago (see vignette 6).

During the Pinedale glaciation, a major glacier occupied the valley dropping away to the northwest. The glacier flowed away from you and to the right, downslope toward Red Lodge. The valley has a U-shaped cross section with steep walls and a bottom that is flat enough to contain a small lake. This U shape is solid evidence that a glacier eroded the landscape. The excavation and removal of bedrock at the base of the glacier produced the flat valley bottom, while erosion of bedrock along the sides of the glacier produced the steep valley walls. In contrast, valleys with bedrock floors that have been eroded solely by a stream or river typically are V-shaped.

You'll notice a number of other U-shaped valleys from your vantage point, all of which were occupied by ice during the Pinedale glaciation. Some of the valley walls have a slightly scooped or bowl-like shape. These bowls, called *cirques*, were excavated by small glaciers called *cirque glaciers*. As the cirque glaciers grew during the Pinedale glaciation, they merged to form the larger valley glaciers that carved out the U-shaped valleys.

There is a sharp ridge just beyond the lake in the valley floor. Called an *arête*, this ridge developed as the glacier occupying the valley in front of you eroded the near side of the ridge, while a second glacier flowing in the same direction eroded the far side of the ridge. These glaciers merged at the downstream end of the ridge. As each glacier eroded the ridge, it became narrower and narrower.

On a clear day, a prominent, pointy peak called Bear's Tooth can be seen on the skyline. It is an especially impressive example of glacial erosion. Sharp peaks like this form when cirque glaciers erode three or more cirques into the sides of a single mountain, creating a steep rock summit called a *horn*.

To the northwest of stop 1, the higher ground above the U-shaped valleys forms a relatively flat surface. In the summer, this high-elevation surface is covered by green tundra that includes a wide variety of small, surprisingly fragrant alpine flowers. The flat surface is also part of the Beartooth Plateau. It was shaped by the ice cap, which completely covered the scene in front of you. The ice removed the layers of sedimentary rock formerly located on top of the crystalline basement rock,

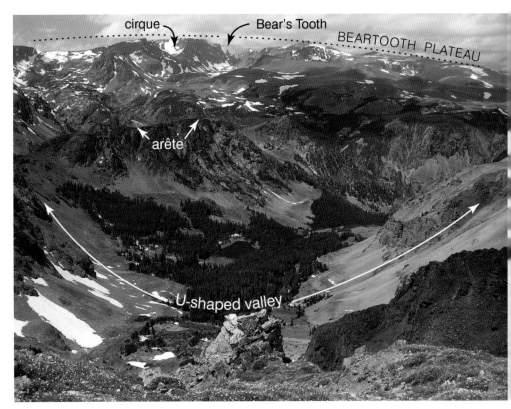

View to the west from Beartooth Pass. Glaciers carved this deep U-shaped valley into Precambrian basement rock of the Beartooth Plateau. The small, pointed Bear's Tooth is a good example of a glacial horn. The dashed line represents the approximate position of the upper surface of the ice cap when it was fully developed.

leaving behind the flat surface. Much of the Beartooth Plateau was covered with sedimentary rock of Paleozoic and Mesozoic age before the ice came.

A few isolated remnants of sedimentary rock were left in places. One of the most spectacular examples is Beartooth Butte, at stop 2. Unlike the massive gray and brown Precambrian basement rock that makes up most of the plateau, Beartooth Butte consists of layered red, yellow, and brown sedimentary rock. Cambrian-age sandstones and limestones are exposed at the base of the butte, a prominent layer of red Devonian-age sandstone forms the middle, and Mississippian-age limestones crop out at the top. Not only did Pinedale glaciers (and probably Bull Lake glaciers) carve the steep walls of Beartooth Butte, they deposited the sediment forming the boulder-strewn, rolling hills at the foot of the butte. These hills are made of glacial till, a jumble of sediment of different sizes and types that was carried by glacial ice and left behind after the ice melted.

Stop 3, the Pilot Peak overlook, is an excellent spot to stop, stretch, and imagine the enormous mass of glacial ice that built up and flowed off the Beartooth Plateau when the Yellowstone Ice Cap was fully developed.

Glaciers carved the steep sides of Beartooth Butte at stop 2 and deposited the rolling hills of glacial till at its base. Beartooth Butte is a remnant of Paleozoic sedimentary rock that remained after glaciers had stripped most of the Beartooth Plateau of this rock.

Looking west from stop 3 on a clear day, it's impossible to miss Pilot Peak: the prominent, isolated mountain with the pointed summit. Like Bear's Tooth, visible from stop 1, Pilot Peak is a great example of a glacial horn, formed as three or more cirque glaciers carved different sides of the mountain. An excellent example of an individual bowl-shaped cirque is visible on the north (right) side of the peak.

When the Yellowstone Ice Cap was at its maximum, its upper surface was about 11,000 feet (3,350 km) above sea level and within about 500 feet (150 m) of the summit of Pilot Peak, with smaller cirque glaciers nibbling away at the sides of the exposed mountaintop. To visualize the Yellowstone Ice Cap, use your imagination to draw a line

View looking west from stop 3. The dashed line indicates the approximate level of the Yellowstone Ice Cap's upper surface during the Pinedale glaciation. At 11,708 feet (3,569 m) above sea level, Pilot Peak rose above the ice surface roughly 500 feet (about 150 m), although cirque glaciers extended upward from the ice cap toward the peak's summit. The rounded hills in the foreground and the mountains on the skyline directly to the left of Pilot Peak were completely covered by the ice cap, whereas the snowcapped peaks of Hurricane Mesa were not. Transparent white arrows denote direction of ice movement.

Telephoto shot of Pilot Peak from stop 3. The steep mountain faces were carved by cirque glaciers. The bowl-shaped north (right) side of the mountain that contains the large snowfield is a great example of a cirque.

PILOT PEAK

corresponding to the ice cap's upper surface. It should start just below the summit of Pilot Peak and project to the left over the top of the closer set of hills but not quite over the tops of the higher mountains (Hurricane Mesa) to the left and farther away.

Geologists have mapped the orientation of scratches and other elongate features the ice left behind on the region's bedrock. These features indicate that the ice cap's summit was about 10 miles (16 km) north of Pilot Peak and about the same distance northwest of stop 3. From where you stand, ice moved from northwest to southeast, or right to left, diverging when it encountered the horn and converging past the horn's south side. Some of the ice spilled eastward and flowed down the long valley that now contains the Clarks Fork Yellowstone River, which is to your south. This ice completely overran all of the low, rolling hills in this valley. In the upper parts of the valley, closer to Pilot Peak, the ice was about 3,500 feet (1,070 m) thick. The ice was more than 1,000 feet (300 m) thick at stop 3. It left behind the large, isolated boulders that are perched on the bedrock surface above the parking area—looking as if they were placed there by a giant hand.

Looking west from the top of the outcrop of basement rock above the parking lot at stop 3. Glacial ice flowing to the southeast shaped this smooth bedrock surface and deposited the boulders when it melted. Pilot Peak is the prominent horn on the skyline.

If you walk up onto the outcrop for a closer look, you'll see that it consists of pinkish granite, which is part of the basement rock forming the Beartooth Plateau. Although the passing glacial ice shaped the sloping bedrock surface, the granite does not display any obvious scratches left by the ice. Weathering has probably removed these rather delicate features. For a look at well-preserved scratches, proceed to stop 4.

Looking at the outcrop at stop 4 from the south side of the road, you'll see that it is composed of two geological units. The east end is Precambrian granite with several smooth faces. Sediment containing rounded boulders and cobbles in a finer-grained matrix of sand and silt is draped on top of this bedrock and forms the west side of the roadcut. This sediment is till that melting glacial ice left behind. As you face the roadcut, glacial ice flowed to the east, down the valley from left to right. Rocks embedded in the ice were dragged against the granite here, leaving behind parallel scratches, called *glacial striations*, on the otherwise smooth surfaces.

Looking northeast at the roadcut exposed at stop 4. Clasts carried by glacial ice left the parallel scratches, or striations, in the bedrock surfaces at the east (right) end of the roadcut, and the faces were polished by fine sediment in the ice. The bedrock is overlain by glacial till that makes up the west (left) end of the roadcut.

If you want, cross the road and take a closer look. Once you've had a chance to inspect the glacial striations, walk west (left) to where the granite is in contact with the overlying glacial till. The poor sorting of this sediment is characteristic of glacial till. Unlike flows of water or wind, which tend to sort and deposit sediments by size, ice moves too slowly for sorting to occur. As a result, when glaciers melt, the sediment they carry is unceremoniously dropped and deposited. There are a variety of rock types in the till, including igneous and metamorphic rocks derived from the Beartooth Plateau; cobbles of limestone (less common) and sandstone (rare) derived from younger rocks of Paleozoic and Mesozoic age; and dark, fine-grained volcanic rocks derived from the Absaroka Volcanic Supergroup (see vignettes 7 and 8).

The assemblage of rock types in the glacial till provides some insight into the wide-reaching erosional effects associated with the buildup and advance of glaciers in Yellowstone. Although glacial erosion profoundly shaped many of Yellowstone's landscapes, much of the specific information about the area's glacial history is preserved in the deposits that the glaciers left behind, including the till at stop 4. In vignette 15, we examine some of these glacial deposits.

Close-up of glacial striations on polished bedrock at stop 4. The glacier flowed from left to right. Rock hammer for scale.

The cobbles and boulders in the glacial till at stop 4 are sedimentary, volcanic, and plutonic. This till sits on top of and next to glacially striated bedrock.

15.
Rivers of Dirty Ice
Glacial Deposits in Northern Yellowstone

When the Yellowstone Ice Cap was fully developed about 17,000 years ago, a hiker would have been able to walk 150 miles (240 km), from the south end of Grand Teton National Park to the north side of the Beartooth Plateau, without stepping on bare rock or solid ground. This hike up and over the Yellowstone Ice Cap would have taken the hiker all the way across Yellowstone National Park. It also would have been possible to walk over glacial ice for 75 miles (120 km) in a westerly direction, starting in the eastern foothills of the Beartooth Mountains, northwest of Cody, Wyoming, and ending up on the western side of the Gallatin Range in northwestern Yellowstone National Park.

From the highest portions of its upper surface, the ice of the massive Yellowstone Ice Cap flowed downhill in all directions toward its margins. The ice was guided by preexisting river valleys down which faster-moving ice, called *ice streams*, flowed within the overall ice cap. The most impressive of these ice streams formed as ice flowing southward off the southern Beartooth Plateau converged with an ice stream flowing generally northward off the northern part of the Yellowstone Plateau. This large ice stream ran generally northwestward down the Yellowstone River drainage system in northern Yellowstone National Park and was the main source for a single, very large glacier that extended into Paradise Valley. Named the Yellowstone Outlet Glacier because it was a major outlet for the ice cap, this glacier was the biggest of numerous outlet glaciers that protruded like fingers from the margins of the ice cap.

Today, many of the mountain fronts in Yellowstone Country contain landforms and sediment deposits related to the maximum advance of glaciers associated with the Yellowstone Ice Cap. Additional landforms and sedimentary deposits related to the retreat, or melting, of the glaciers and the ultimate demise of the Yellowstone Ice Cap are preserved within mountain ranges and some of the broader valleys between them. In this vignette, we examine glacial landforms and deposits related to both the buildup and disappearance of the ice cap.

Glaciers are effective agents of erosion (see the discussion of the "glacial buzz saw" theory in vignette 14), producing not only the characteristic landforms discussed in vignette 14 but large volumes of rock debris that the ice carries downstream. When the rock debris reaches the end of a glacier or is pushed off to the sides of a glacier, it typically accumulates as a bermlike feature called a *moraine*. The rock debris in all moraines, called *till*, is a jumbled mixture of sediment containing clasts that range from clay sized to boulder sized. A moraine that forms at the downstream end, or snout, of a moving glacier is called an *end moraine*. The end moraine that is left where a glacier reaches its position of maximum advance is called a *terminal moraine*; these tend to be best developed when a glacier remains in its terminal position for decades or centuries, allowing a great deal of till to pile up before the glacier retreats. Till that builds up along the sides of a glacier forms a *lateral moraine*. In addition, significant amounts of till can accumulate at the base of a glacier, forming what's called a *ground moraine*. When a glacier finally melts and retreats back upslope, the terminal and lateral moraines that are left behind provide an outline of the glacier when it was fully developed.

Exceptionally well-preserved moraines can be seen in the Pine Creek valley at stop 1. When the Yellowstone Ice Cap was at its fullest, ice was funneled into the eastern Paradise Valley, which stretches between Livingston and Gardiner, through canyons and valleys like the Pine Creek valley. These ice streams formed valley glaciers. As you approach the mouth of the valley from the west, you'll notice a set of low, rolling hills well below the rugged Absaroka Range forming the skyline. These hills are terminal and lateral moraines left by the Pine Creek valley glacier. Using cosmogenic surface exposure dating techniques to determine the age of boulders in the moraines, geologists have concluded that this glacier was at its maximum advance about 17,500 years ago. The valley beyond the moraines has an obvious U shape, having been carved by the valley glacier. When the moraines were deposited at the mouth of the valley, ice stretched from there southeastward for 70 miles (113 km), up and over the Absaroka and Beartooth ranges.

GETTING THERE

Diverse glacial deposits can be viewed in the northern Yellowstone region between Livingston, Montana, and the Tower Junction area in Yellowstone National Park. To get to stop 1 from Livingston, head 3.2 miles (5.1 km) south of the I-90 overpass to the intersection of US 89 and East River Road (Montana 540). Turn east (left) onto Montana 540 and proceed 7.7 miles (12.4 km) to the intersection with Luccock Park Road. Turn east (left) on Luccock Park Road and continue for 1.9 miles (3.1 km). The road switches back a couple of times as it climbs up the front of the Pine Creek valley terminal moraine. Stop 1 is the small roadside pullout on the south side of the road just beyond the top of the climb. To get to stop 1 from Gardiner, Montana, start at the bridge over the Yellowstone River and proceed north on US 89 for 19.5 miles (31.4 km). Make a slight right-hand turn onto East River Road (Montana 540) and proceed another 24.1 miles (38.8 km) to Luccock Park Road. Turn right and continue 1.9 miles (3.1 km) to the roadside pullout. To reach stop 2, backtrack to Montana 540 and head south 11.4 miles (18.3 km) to a small pullout on the west (right) side of the road (1.2 miles, or 1.9 km, past the small settlement of Pray). Please be very careful at this stop as the traffic typically moves fast along this stretch of highway. To reach stop 3, proceed south on Montana 540 for 1.3 miles (2.1 km) to the intersection with Murphy Lane. Turn right on Murphy Lane and continue for 1.2 (1.9 km) miles to Emigrant. Turn south (left) on US 89 and continue 36.5 miles (58.7 km) to Mammoth Hot Springs. From Mammoth Hot Springs, turn east (left) onto Grand Loop Road and continue for 18.1 miles (29.1 km) to Tower Junction. Turn left there, heading toward Northeast Entrance, and proceed 3.2 miles (5.1 km) to a small roadside pullout on the north (left) side of the road, stop 3. There is a large boulder at the east end of the pullout.

Looking northeast from stop 1, you can't help but notice the contrast between the rugged mountain peaks of the Absaroka Range on the skyline and the low ridgeline directly across the valley. This low ridgeline is the lateral moraine that formed on the north side of the Pine Creek valley glacier, which flowed almost directly west (right to left) out of the Pine Creek valley and about 1 mile (1.6 km) beyond stop 1 into Paradise Valley. The main road at stop 1 is straight because

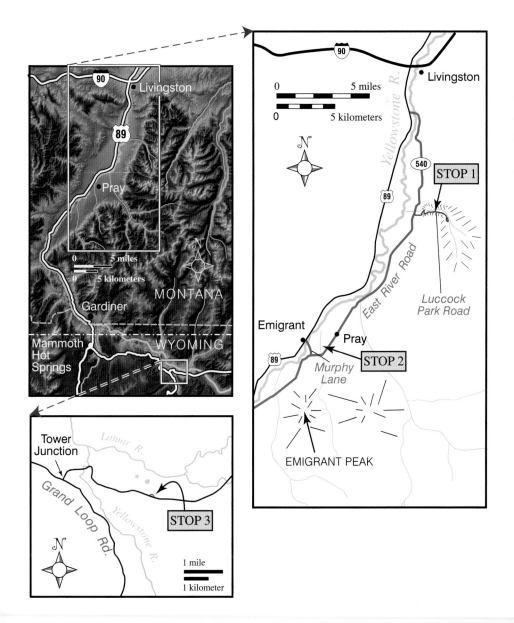

it follows a similar, parallel ridge comprising the lateral moraine that formed on the south side of the valley glacier. In fact, you are standing on this moraine. When the glacier was depositing these moraines, ice stretched from one lateral moraine to the other.

The many boulders strewn about the two moraines were delivered by the Pine Creek valley glacier. Virtually all of the boulders consist of Precambrian basement rock derived from outcrops in the headwaters of Pine Creek. Most of the boulders are granite or gneiss, and many of the gneiss boulders display good foliation, mineral layering that developed due to the high temperatures and pressures of metamorphism deep within the Earth.

Although the Pine Creek valley glacier transported numerous large boulders, it also carried finer sediment, including sand, silt, and clay. Most of these finer sediments developed as rocks embedded within the glacier were slowly ground together. The finest sediment is called *glacial flour*. Collectively, this material can cause glacial ice to be surprisingly dirty stuff, particularly near the terminus of a glacier, where the debris and sediment collected across much of the glacier's expanse are concentrated.

Looking northeast across the Pine Creek valley at stop 1. Both lateral moraines of the Pine Creek valley glacier are visible.

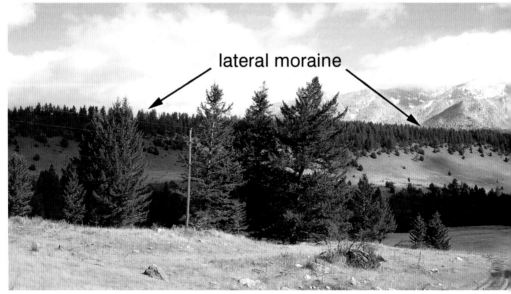

Unlike the two sharp-crested lateral moraines, the terminal moraine of the Pine Creek valley glacier is a series of rolling, boulder-strewn hills that form a broad arc between the two lateral moraines. Luccock Park Road passes over the terminal moraine as it switches back and forth several times between stop 1 and the floor of the Paradise Valley, where the road heads straight west.

When the Pine Creek valley glacier was at its full advance, meaning it had reached its greatest extent, its terminus was located less than 10 miles (16 km) from the terminus of the Yellowstone Outlet Glacier. When fully formed, the Yellowstone Outlet Glacier was probably the longest single glacier that grew from ice entirely within the contiguous United States. The most far-traveled molecules of ice within the glacier flowed more than 85 miles (137 km) from the summit region of the ice cap near the northeastern boundary of Yellowstone National Park. From the Gardiner area, the glacier flowed northwest through Yankee Jim Canyon before turning nearly 90 degrees to the right where Tom Miner Creek flows into the Yellowstone River. From there, the glacier advanced an additional 20 miles (32 km) to the northeast, flowing over the white Eocene siltstones at Hepburn's Mesa (see vignette 10) and past the future locations of Chico, Emigrant, and Pray.

lateral moraine

A boulder of Precambrian gneiss in the terminal moraine of the Pine Creek valley glacier near stop 1. Nearly all the boulders in the Pine Creek valley moraines consist of igneous and metamorphic basement rock eroded from the Absaroka and Beartooth ranges.

On your way to stop 2, you cross the outlet glacier's terminal moraine, called the Eightmile Creek moraine, about 3 miles (4.8 km) north of Pray. It is not very accessible on public land. Stop 2, however, provides a great view of a well-preserved recessional moraine of the outlet glacier and presents an interesting comparison with the relatively small moraines of the Pine Creek valley. Recessional moraines form when a retreating glacier temporarily stabilizes for a few decades or centuries, depositing till in the same spot. Although a couple of private driveways and ranch roads turn off East River Road (Montana 540) in the vicinity of stop 2, it's best to pull off on the shoulder of the road and be mindful of the traffic while taking in the geology.

The majestic peak to the south is Emigrant Peak. Directly below the peak's summit is an obvious steep-sided valley that separates Emigrant Peak from the high, timbered mountains to the northeast (left and toward you). This is Emigrant Gulch. During the Pinedale glaciation, a small valley glacier occupied this valley and flowed to the west, from left to right. A dozen or more cirque glaciers located up the drainage and as far away as Mineral Mountain, about 9 miles (14.5 km) southeast of here in the heart of the Absaroka-Beartooth Wilderness, provided the Emigrant Gulch valley glacier with ice. This glacier joined the Yellowstone Outlet Glacier at the mouth of the canyon, where the settlement of Old Chico is located.

Take a look at the northwestern (right) side of the partially timbered ridge in front of Emigrant Peak. By mapping glacially deposited boulders on this side of the ridge, geologists with the U.S. Geological Survey determined that the top of the Yellowstone Outlet Glacier was 6,500 feet (1,980 m) above sea level there. The upper limit that glacial ice reaches on a mountainside is known as its *trim line*. In this case, the trim line is roughly one-third of the way up the west side of the ridge, at the point where the rolling hills end and the side of the ridge becomes steeper. Imagine how the Yellowstone Outlet Glacier looked about 16,500 years ago when it had reached this trim line. The glacier stretched from there to the west, completely across Paradise Valley to the hills more than 10 miles (16 km) away. Stop 2 was under approximately 1,000 feet (300 m) of ice and located roughly 5 miles (8 km) south of the glacier's terminus.

The lower, grass-covered hills between you and Emigrant Peak belong to the Chico moraine, a recessional moraine the Yellowstone Outlet Glacier deposited about 16,100 years ago when it paused during its retreat. The ice forming the margin of the glacier would have been visible from stop 2, but it probably looked less like ice and more like an extension of the moraine because of all the material the ice was carrying.

Glaciologists studying the dynamics of glaciers have found it useful to define a feature called the *equilibrium line altitude* (ELA), which is the elevation separating the zone where enough snow accumulates

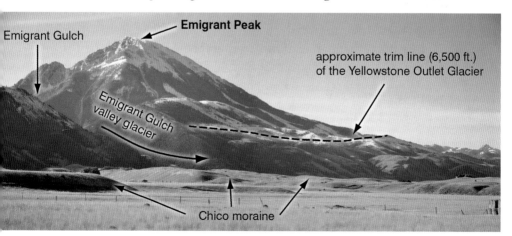

Looking south from stop 2. During the Pinedale glaciation, a valley glacier occupied Emigrant Gulch and merged with the Yellowstone Outlet Glacier at the mouth of the gulch. The Chico moraine is a major recessional moraine deposited by the Yellowstone Outlet Glacier.

in mountainous terrain to form glacial ice from the zone where not enough accumulates. Any glaciers that advance downslope below the ELA must continually be fed by glacial ice forming at higher elevation. The ELA is not fixed; it will shift downslope as the climate cools and upslope as it warms. At the peak of a glacial period, when the glaciers are most fully developed, the ELA will be at its lowest elevation.

When the Yellowstone Outlet Glacier was fully advanced, the ELA was located approximately at Gardiner. All of the ice that flowed past Gardiner had formed at higher elevations within the Yellowstone Ice Cap. The area from which the glacier's ice came, called the *ice collection zone*, was defined on its upstream end by an ice divide—a broad topographic ridge that formed on the upper surface of the ice cap. Ice divides are similar to topographic divides in that ice on the opposite sides of an ice divide flows in opposite directions, just as water on the opposite sides of a topographic divide flows in opposite directions. Ice divides commonly coincide approximately with topographic divides beneath them, particularly in mountainous regions. During the Pinedale glaciation, a major ice divide developed around the Lamar River drainage and tributaries in and around northern Yellowstone National Park. The ice divide had a squashed semicircular shape, and ice that flowed off the divide into the interior of the semicircle was ultimately funneled down the Yellowstone River drainage and into the Yellowstone Outlet Glacier.

Ice within the outlet glacier's collection zone came from five different regions. The most significant contributions came from the southern Beartooth Plateau; ice spilled southward off the plateau before curving west and continuing down the Lamar River valley. The most far-traveled ice came from the high country forming the headwaters of the Lamar River and flowed mostly northwest. Significant amounts of ice also flowed northward off the Yellowstone Plateau, much of it coming from the northern Washburn Range, the Obsidian Creek drainage south of Bunsen Peak, and the northern and eastern Gallatin Range.

As the climate began to warm following peak glacial conditions, the ELA began to rise in elevation. As the ELA climbed and the zone of accumulation within the Yellowstone Ice Cap shrank, less glacial ice was funneled downslope, causing outlet glaciers along the margins of the ice cap to retreat. As melting progressed, the ice cap disintegrated into a series of smaller masses of ice situated over the higher regions in and around northern Yellowstone National Park. The biggest mass was an ice cap over the Beartooth Plateau. Ice on the Yellowstone Plateau, which is at a lower overall elevation than the Beartooth Plateau, melted back to a few relatively small masses centered over the Gallatin Range, Washburn Range, and parts of the eastern Absaroka Range.

As the Yellowstone Ice Cap disintegrated, the dynamics of the flowing ice changed. Among the most significant changes were those that occurred within the Lamar River valley. As noted before, during peak glacial conditions the Lamar River drainage captured ice flowing northward off the Yellowstone Plateau and southward off the Beartooth Plateau. When the ice began to melt and the ELA began to climb, ice flowing northward off the Yellowstone Plateau retreated first because of its lower elevation. In contrast, the higher Beartooth Plateau continued to generate large amounts of ice, much of which flowed to the south. The ice coming off the Beartooth Plateau was so voluminous that it overwhelmed the remnants of the retreating Yellowstone Plateau ice. It pushed southward well past the Lamar River and actually flowed up the opposite side of the Lamar River valley. As deduced from the position of boulders carried into the valley from the Beartooth Plateau, this ice gained as much as 1,600 feet (490 m) of elevation and reached a terminal position nearly 10 miles (16 km) beyond the floor of the Lamar River valley.

As climate continued to warm at the end of the Pinedale glaciation, the mass of ice on the Beartooth Plateau disintegrated into isolated, southwestward-flowing valley glaciers that moved down Slough Creek and Soda Butte Creek in northeastern Yellowstone National Park. An especially well-preserved ground moraine left behind by the Slough Creek glacier can be seen at stop 3.

A ground moraine is an irregular blanket of till deposited at the base of a glacier as sediment carried within the ice becomes lodged against the ground surface or is simply dropped as the glacier retreats. The ground moraine is well preserved here because it was not overridden by younger ice. The best view is to the north across the floor of the Lamar River valley. The rolling hills, which are sometimes described as "hummocky," are characteristic of ground moraines. They are strewn with large boulders carried here by the ice. Small bogs and ponds are common in the ground moraine. Some of these formed as a large mass of ice broke off the front of the retreating glacier and eventually melted. The roughly circular depressions that these large masses of ice left behind when they melted are called *kettles*.

If you walk a short distance out onto the ground moraine, you'll see that virtually all of the boulders are made of igneous or metamorphic rock of the Precambrian basement. Although Precambrian basement rock is widely exposed across the Beartooth Plateau north of stop 3 and even forms some of the high hills on the north side of the Lamar River valley, this rock is almost nonexistent at the surface south of the Lamar River. With the exception of one outcrop on the north end of Specimen Ridge, Precambrian crystalline rock has not been mapped anywhere south of Northeast Entrance Road in all of Yellowstone National Park. Clearly, the abundant boulders in the ground moraine came from the north and were carried here by glaciers.

As the residual ice masses on the Beartooth and Yellowstone plateaus continued to melt, large volumes of water were produced. Called *meltwater*, much of it found its way to the base of the remaining glaciers, where it acted as a lubricant and caused the flowing ice to reshape previously deposited glacial till. In some spots, glaciers temporarily dammed meltwater, forming glacial lakes. As the glaciers continued to retreat these ice dams failed, releasing large floods. We'll turn to these topics in vignette 16.

The rolling hills north of stop 3 are composed of ground moraine—till dropped by the last glaciers to retreat from the Lamar River valley. Numerous boulders derived from the Beartooth Plateau are visible in the foreground.

16.
Melting Ice and Sliding Shale
Floods and Earthflows near Gardiner

The warming climate at the end of the Pinedale glaciation caused melting across much of the Yellowstone Ice Cap and greatly decreased the volume of glacial ice being generated at higher elevations. As less ice flowed downward toward the margins of the ice cap, its outlet glaciers retreated rapidly. The record of glacial retreat is perhaps best known for the Yellowstone Outlet Glacier, discussed in vignette 15. Using cosmogenic surface exposure dating, geologists have determined that boulders in its terminal moraines were deposited, on average, about 16,500 years ago, indicating that the outlet glacier began retreating after this time. It paused in its retreat to deposit the recessional Chico moraine 16,100 years ago. Then, by about 14,200 years ago, the front of the outlet glacier had withdrawn to the Gardiner area—about 36 miles (58 km) south of the glacier's terminal position.

Rapid melting of the Yellowstone Ice Cap produced large volumes of meltwater, much of which worked its way down to the ground through passageways in the ice. Once on the ground it flowed beneath the ice, moving generally downslope and ultimately emerging from the downstream end of the Yellowstone Outlet Glacier as a cold, fast-flowing, muddy river. Although most of the meltwater flowed away from the outlet glacier in this manner, significant amounts were trapped in ice-dammed glacial lakes.

Today, ice-dammed glacial lakes are very common and form in several different ways. Some form when a valley glacier moving down a tributary valley advances across and dams a stream in the main valley. Others form when a glacier advances across a water-filled fjord—a long, narrow inlet filled with marine or lake water. The glacier isolates the water in the head of the fjord, forming a lake. Lakes can also develop under ice. These subglacial lakes form when heat from the Earth causes the base of a large ice cap to melt. The ice above the lake insulates the water from the cold atmosphere and forms a hydrostatic seal between the lake and the ice. The weight of the ice and the presence of the seal cause the lake water to be pressurized, and the seal keeps the water from leaking upward or otherwise escaping. The high pressure lowers the temperature needed to freeze the water in the lake. In 2009, geoscientists used radar technology to detect 124 subglacial lakes under the Antarctic Ice Cap. Some are situated over areas of geothermal activity, and many appear to be interconnected by flowing water.

GETTING THERE

Impressive deposits of glacial outburst floods and massive earthflows can be seen in northern Yellowstone Country. Stop 1 is 49.4 miles (79.5 km) south of the I-90 overpass at Livingston and 3.5 miles (5.6 km) north of the bridge over the Yellowstone River in Gardiner. Park along the west shoulder of US 89 and proceed on foot to the fence about 3 feet (0.9 m) off the road. Stop 2 is 0.8 mile (1.3 km) south of stop 1, also along the shoulder of US 89. Please exercise extreme caution at both stops as the traffic here moves fast. To examine the flood bar seen from stop 1 up close, continue to stop 3. Follow US 89 through Gardiner. Just before the original stone archway marking the North Entrance to the park, turn northeast (right) onto Old Yellowstone Trail South. Drive 3.5 miles (5.6 km), passing the Gardiner High School, to a place where the road curves sharply to the right around the boulder-strewn flood bar. Park along the road. A fantastic view of a drumlin field can be had from the summit of Bunsen Peak. To reach the Bunsen Peak Trailhead, proceed 5.3 miles (8.5 km) south of the stone archway to the intersection of North Entrance Road and Grand Loop Road in Mammoth Hot Springs. Head toward Norris on Grand Loop Road for 4.8 miles (7.7 km) to the trailhead parking lot on the east (left) side of the road. From the parking lot, the trail climbs about 1,300 feet (400 m) and spans a horizontal distance of 2 miles (3.2 km) to the summit. The round-trip hike will take an experienced hiker 3 to 4 hours and is moderately strenuous. Expect snow near the summit before mid-June.

As the Yellowstone Ice Cap disintegrated, several glacial lakes of substantial size formed in the headwaters of the Lamar River. One lake temporarily occupied part of the upper Lamar River when it became blocked by ice flowing southwest off the Beartooth Plateau. A second lake backed up into the Grand Canyon of the Yellowstone when glacial ice, also coming off the Beartooth Plateau, crossed the confluence of the Lamar and Yellowstone rivers. Subglacial lakes almost certainly developed at the base of the ice cap above major thermal areas associated with the Yellowstone Volcano, although geologists have not established the size and dimensions of these lakes with much certainty.

Glacial lakes are unstable features because the ice that impounds them is temporary. Not only does ice float, but heat from the relatively warmer lake water can melt it, decreasing the size and effectiveness of a glacial dam. Eventually, impounded water works its way around or

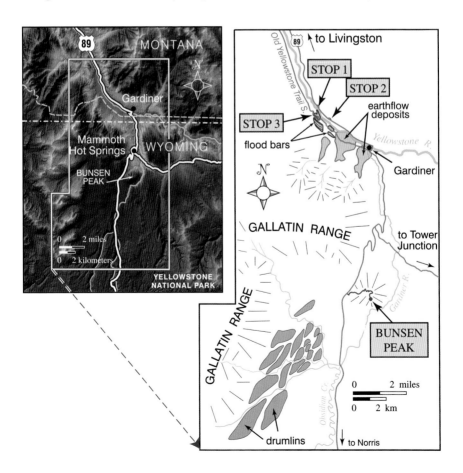

under the ice, rapidly enlarging passageways through which it travels and weakening the dam in the process. Glacial dams can fail catastrophically, releasing the entire contents of a lake in as little as a few minutes. These large floods of meltwater are called *glacial outburst floods*.

Glacial outburst floods are by far the largest floods known to occur on Earth. In 1986, a glacial outburst flood observed in Alaska involved an estimated 35 million cubic feet (1 million cubic meters) of water per second—more than thirty times the volume of water flowing out of the Mississippi River when it is flooding and about ten times the amount released by the Amazon River, which delivers about one-fifth of the world's freshwater to the oceans. Although the flood was impressive, prehistoric glacial outburst floods recognized in the geologic record were substantially larger. The most famous of these occurred at the end of the Pinedale glaciation, when a large valley glacier in northern Idaho crossed the Clark Fork River and impounded Glacial Lake Missoula in western Montana. The largest flood that occurred when this ice dam failed is estimated to have been about ten times the combined flow of all the rivers in the world!

Ice was not the only thing that impounded large, temporary bodies of water. As the ice retreated, the ground surface over large parts of Yellowstone Country reemerged. Where the recently thawed ground was steep, saturated with water, and underlain by fine-grained sedimentary rock, it was unstable and prone to massive downslope movement—events called *earthflows*. Large earthflows blocked the Yellowstone River in several places, and each mass had the potential to dam the river. Like the ice dams, these earthen dams would have been unstable and may have released catastrophic floods.

At least two very large floods rushed down the Yellowstone River valley, modifying its shape as they went. We know they occurred after the Yellowstone Outlet Glacier retreated from the Gardiner area, placing them sometime in the past 14,200 years. In some spots the floodwaters scoured deep pockets into hard bedrock; elsewhere they left large, streamlined deposits of boulders. In this vignette, we examine the evidence for large floods at three locations below Gardiner and visit one earthflow deposit.

Much of the meltwater produced as the Yellowstone Ice Cap melted flowed down the Yellowstone River valley, through Yankee Jim Canyon and the Paradise Valley. As the exceptionally strong glacial outburst floods flowed over recently deposited till, they were powerful enough to pick up boulders, transporting and organizing them into large land-forms called *flood bars*. Although flows of this magnitude are rarely observed in nature, geologists know something about how flood bars form by having studied similar but smaller accumulations of finer sediments. Generically, these sorts of deposits are known as *sediment bars*, no matter what the size of the sediments. Sediment bars form when moving water picks up sediment and later deposits it as the flow loses strength. (Most of Yellowstone's flood bars occur where the floodwaters would have spread out and slowed down below a constriction in a valley.) Sediment bars form in all sorts of modern environments and in response to many different kinds of flows, including those associated with rivers, tides, or other currents in oceans or lakes. These bars also vary tremendously in size, ranging from river bars smaller than 1 to 2 feet (30 to 60 cm) high and about 10 yards (9 m) across to ocean tide bars, which can be more than 100 miles (160 km) long, 6 miles (10 km) wide, and 130 feet (40 m) high.

Although many flood bars occur in close proximity to a modern, flowing stream, they typically exist at elevations tens or hundreds of feet (up to about 100 m) higher than the modern stream and hundreds or thousands of feet (1 km or more) away from the stream channel. These spatial relationships to the modern stream indicate that the bar was deposited by a flow that was much larger than the modern stream. In addition, flood bars typically are much larger than the sediment bars forming in the modern stream and contain large boulders that only especially powerful flows could transport.

Perhaps the single best example of a flood bar in Yellowstone Country can be seen at stop 1. One or more glacial outburst floods deposited it about 13,700 years ago. From stop 1, look south-southeast just beyond the Yellowstone River. The bar is the pair of low, boulder-strewn ridges with an old railroad cut through it. The crest of the flood bar, which had to have been covered by water during flooding, is about 40 feet (12 m) higher than its edges, indicating that the floodwaters were at least this deep in places. The crest is also 60 feet (18 m) higher than the modern Yellowstone River. That means that during flooding the ancestral Yellowstone River must have swelled to about 0.5 mile

(0.8 km) across here, reaching approximately from US 89 west to the base of the higher hills beyond the flood bar.

The grass-covered hills on the far side of the valley, above, beyond, and slightly left of the flood bar, were deposited by a large earthflow. An earthflow is a large, viscous flow of water-saturated—or mostly saturated—sediment that moves downslope due to gravity and is characterized by significant internal deformation. The behavior of earthflows is between that of slumps, which tend to move downslope as a single intact mass sliding along its base, and mudflows, which move downslope quickly and with a great deal of internal deformation. Earthflows typically occur in places where clay-rich sediment is located on a slope. Because the degree to which a mass is saturated with water directly affects its ability to move, the speed at which earthflows travel can vary greatly. Relatively dry earthflows move less than 1 foot (30 cm) per year, while saturated earthflows are capable of rushing downslope at 35 miles per hour (56 km per hour). When they stop flowing, earthflows usually come to rest in a lobe-shaped deposit. For a better perspective on the earthflow deposit and a great view up the Stephens Creek valley, the source of the earthflow, continue to stop 2.

From stop 2, the rumpled hills directly across the Yellowstone River valley, below the evergreen and aspen trees at the mouth of the Stephens Creek valley, comprise the earthflow deposit. The perimeter of the deposit occurs where the rumpled hills meet the much flatter ground, about 0.5 mile (0.8 km) short of the Yellowstone River. The Stephens Creek earthflow was one of several that occurred in this area. The earthflows originated in outcrops of fine-grained, Cretaceous-age

Looking south-southeast from stop 1. The flood bar comprises the pair of boulder-covered ridges in the center of the photograph.

shale in the mountains flanking both sides of the Yellowstone and Gardner river valleys. Most of the flows came from the west sides of the valleys because the layered sedimentary rocks there are tilted gently to the east, toward the valley floors. The earthflows started when shale became saturated with water, and the slippage probably began along one or more bedding planes (visible layering in the rock) within the shale. Rocks on the east sides of the valleys are less susceptible to slippage because they tilt away from the valley floors.

The Stephens Creek earthflow was a relatively modest event in the Yellowstone River valley. If you turn about 90 degrees to the left and look southeast, up the Yellowstone River valley, you'll notice that the low hills that reach all the way to the river have the same rumpled look as the upper surface of the Stephens Creek earthflow deposit. In fact, nearly all of the low-elevation, grass-covered hills to the southeast and left of the Stephens Creek earthflow belong to a much larger earthflow that moved down Landslide Creek. This deposit temporarily blocked the Yellowstone River less than 0.5 mile (0.8 km) north of the outskirts of Gardiner and likely impounded a temporary lake. Since no sedimentary record of this lake has been reported, it may have been relatively small or perhaps existed for only a short time.

Although the earthflow deposits are large features that dominate much of the landscape around the park's North Entrance, their exact age has not been well established. However, the fact that the Landslide Creek earthflow partially covers bouldery outburst flood deposits indicates that it occurred after those floods. Similarly, the Stephens Creek earthflow is not cut by any outburst flood deposits, suggesting

The Stephens Creek earthflow as seen from stop 2.

that it occurred after the outburst floods ripped through the region or that none of the floods reached high-enough elevations to erode it. Most likely, the earthflows occurred shortly after deglaciation, when the permafrost in the Cretaceous shale of the valley walls began to thaw. Shale with thawing permafrost is particularly susceptible to earthflows because as the shale near the surface becomes saturated with water, it simply slides off the shale beneath that still contains permafrost.

From stop 2, it is worth taking an extra ten minutes or so to proceed to stop 3, located on the western edge of the well-formed flood bar seen from stop 1. The route between the two stops takes you over the Landslide Creek earthflow deposit, past boulder-strewn outburst flood deposits, and by additional views of the Stephens Creek earthflow deposit.

From stop 3, the most obvious feature of the flood bar is the multitude of large boulders, the largest of which exceed 6 feet (1.8 m) in diameter. Most of the boulders are well rounded, suggesting they were transported by water. Almost without exception, the boulders are crystalline Precambrian basement rock derived from the Beartooth Uplift. Although younger sedimentary and volcanic rocks crop out widely in this area, they are generally softer and don't form boulders that can withstand much transport. In contrast, boulders of the much harder igneous and metamorphic Precambrian rocks can withstand transport for more than 100 miles (160 km). Floodwaters probably worked most of the boulders out of older glacial moraines along the flanks of the Yellowstone River valley and tributary valleys upstream.

Cosmogenic surface exposure dating of boulders within the flood bar indicates that it formed about 13,700 years ago. That means the flood bar was deposited here a few hundred years after the snout of the retreating Yellowstone Outlet Glacier passed through this area but before ice had completely left the Yellowstone River valley farther upstream.

If you wander around the surface of the flood bar, you'll notice several regularly spaced, elongate, bouldery ridges separated by swales. The ridges and swales are most prominent along the western and northern parts of the flood bar. The ridge crests are up to 8 feet (2.4 m) higher than the intervening swales and are separated from each other by about 10 to 25 yards (9 to 23 m). U.S. Geological Survey geologists have interpreted these features as giant ripples. Although many giant ripples have been described elsewhere and are often cited as evidence of large glacial outburst floods, it is not well understood how they form in moving water. Unlike smaller ripples, which can be studied in natural settings and laboratory experiments, giant ripples do not form very often and are extremely difficult to reproduce in a laboratory.

Stop 3 is a good place to ponder the many ways this segment of the Yellowstone River valley has changed since the snout of the Yellowstone Outlet Glacier retreated south, passing this spot about 14,200 years ago. A relatively large, ancestral Yellowstone River ran through the valley, carrying brown water heavily laden with suspended sediment. Within a few hundred years of the retreat of the ice, the river swelled abruptly, spreading all the way across the valley and inundating it with roaring floodwaters that were likely more than 100 feet (30 m) deep in places. At least two of these large floods ripped through the Yellowstone River valley about 13,700 years ago. Perhaps a few hundred or a thousand years later, large earthflows oozed eastward out of

Looking southwest at the glacial outburst flood bar at stop 3. The rounded boulders are almost exclusively crystalline Precambrian basement rock. The northern and eastern parts of the flood bar are marked by ridges and swales that geologists have interpreted as giant ripples. One of the giant ripples is in the foreground, on the right side of the photo, and the man is standing on the other.

some of the tributary valleys, partially covering the older flood deposits. All of these scenes underscore the tremendously dynamic nature of the geology of the Yellowstone region.

Bunsen Peak

For those inclined to further investigate the ways in which the landscapes of northern Yellowstone National Park changed due to the interaction of ice, water, and sediment at the end of the Pinedale glaciation, I highly recommend hiking to the top of Bunsen Peak. If you drive to the Bunsen Peak Trailhead from stop 3, you'll see several other large earthflow deposits on the west (right) side of the road after passing through the stone arch at the park's North Entrance.

At the summit of Bunsen Peak, try using your imagination to visualize the Yellowstone Ice Cap as it must have appeared when it was fully developed. The summit was under about 600 feet (180 m) of ice. The ice surface rose to the south before reaching an ice divide about 10 miles (16 km) away. The ice flowed generally north from the divide, over and around Bunsen Peak. Ice from the Gallatin Range to the southwest and the Washburn Range to the southeast converged as it approached Bunsen Peak, and all of it was funneled into the Yellowstone Outlet Glacier (see vignette 15).

Look for isolated boulders, mostly dark basalt, around the summit. These contrast markedly with the lighter-colored Eocene volcanic rock making up the peak itself. Bunsen Peak is one of several volcanoes associated with the Absaroka Volcanic Supergroup (see vignettes 7 and 8) and is about 48 million years old. The basalt boulders were derived from the Swan Lake Flat Basalt, which underlies the low-lying region immediately south of the peak. The basalt is between 640,000 and 320,000 years old. Ice transported the boulders to the summit, carrying them upward nearly 1,000 feet (300 m) from their original location.

Although the summit provides a fantastic vantage point from which to imagine the ice cap, it also provides a good opportunity to see a glacial landform that some geologists think is produced by meltwater. If you look southwest to the area between circular Swan Lake and the rugged Gallatin Range, you'll see a series of low, elongate hills oriented parallel to the front of the range. These hills are called *drumlins*, from the Gaelic *druim*, meaning "crest of the hill."

The most interesting thing about the field of drumlins is that their long axes do not parallel the major U-shaped valleys of the Gallatin Range; rather, the axes parallel the front of the range itself. This relationship indicates that the drumlins formed under the ice cap when it stretched completely across the Obsidian Creek valley below you, rather than at a time when the only glaciers in the Obsidian Creek valley were valley glaciers protruding from the U-shaped valleys of the Gallatin Range. The ice of the ice cap flowed north-northeast toward the Yellowstone Outlet Glacier, cutting across the east-facing drainages of the Gallatin Range while shaping the drumlins. Drumlins were first described in the early nineteenth century, but how exactly these landforms are created at the base of glaciers is still a controversial topic. Two main competing theories exist. The first theory posits that drumlins form in areas composed of well-drained sediment or bedrock, which serves as a core over which water-saturated, deformable sediment is accreted as the ice flows over it. The ice shapes the sediment into elongated hills that parallel the direction the ice flows. During deglaciation, more water would have been available to saturate the sediment and may have facilitated the formation of the drumlins.

More controversial is the theory that large floods of meltwater form drumlins. As mentioned before, meltwater can accumulate as subglacial lakes in areas where heat melts the ice at the base of an ice cap. This second theory posits that eventually enough water collects that it is able to lift the ice cap off its bed, allowing vast, turbulent sheets of water to surge outward, scouring and sculpting drumlins from sediment that exists below the ice cap.

If indeed meltwater temporarily lifted the ice cap from its bed and sculpted the drumlins in the Obsidian Creek valley, might there be a connection between the drumlins and the glacial outburst floods that rushed down the Yellowstone River valley? No one knows the answer, in part because no one knows for sure how drumlins form, but also because the detailed geologic history of northern Yellowstone Country is still being deciphered. With the work of future geologists, however, one day we will no doubt know more about the fascinating processes and sequence of events that led to the landforms and sedimentary deposits in this region.

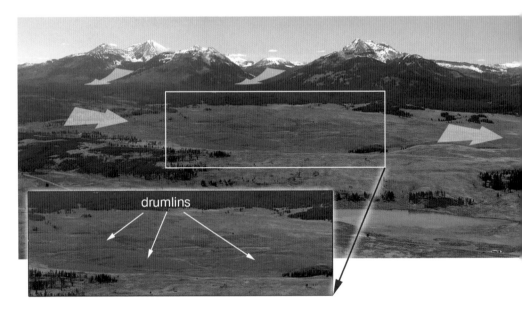

Looking southwest from the summit of Bunsen Peak. Ice carved the sharp peaks and broad U-shaped valleys of the Gallatin Range in the distance. When the Yellowstone Ice Cap was fully developed, ice covered all but the highest peaks and flowed from south to north (left to right). Transparent white arrows denote direction of ice movement. Slightly elongated drumlins (mounds) are visible in the center of the photograph and highlighted in the telephoto inset.

17.
Terraced Travertines
Mammoth's Famous Hot Springs

Nowhere else on Earth does there exist a collection of thermal features as numerous and diverse as those in Yellowstone National Park—more than ten thousand according to the U.S. Geological Survey. These features account for much of Yellowstone's fame as a place of natural wonder and beauty and are the main attraction for most of the millions of people who visit the park each year. In this vignette, we visit the most extensive set of actively forming hot springs terraces in the Yellowstone region at Mammoth Hot Springs, and we'll also look at some of the rock deposited by ancient terraced hot springs.

Thermal features, including hot springs, mud pots, geysers, and fumaroles, are holes in Earth's crust from which hot fluids escape, including water, steam, gases, and vapors. Each type of thermal feature has an opening at the surface called a *vent*, which is connected to a deeper network of openings in the subsurface. Most of the water in Yellowstone's thermal features comes from rain and snowmelt that percolates into the ground, in some cases reaching depths of more than 1 mile (1.6 km). Hot bedrock in the subsurface heats the water, lowering its density and causing it to circulate back upward to thermal features at the surface. The bedrock is heated both by the thermal plume beneath Yellowstone and by rhyolite magma that exists as little as 2 miles (3.2 km) below the surface (see vignettes 11 and 13).

On its subterranean voyage, the thermal water picks up various gases, including carbon dioxide, hydrogen sulfide, argon, and helium. These gases are discharged from the thermal features along with the water. Interestingly, some of the helium is a relatively rare form (on

Earth at least) called helium-3. All helium-3 on Earth formed in nuclear fusion reactions that took place in the center of ancient stars similar to our sun. The explosions of these stars at the end of their life cycle, billions of years before Earth existed, spread the helium-3 across outer space. It mixed with the dust and other cosmic matter that formed the Earth about 4.6 billion years ago. Earth's mantle still contains a significant amount of helium-3, which is carried upward by the thermal plume and circulating thermal water beneath Yellowstone.

Many regard the marvelous hot springs terraces at Mammoth as the park's most beautiful feature. In fact, striking photographs and paintings of the Mammoth area from the 1871 Hayden Expedition helped convince the U.S. Congress to create Yellowstone National Park in 1872. The terraces are made of travertine, a type of freshwater limestone that springwater precipitates, or deposits, through chemical reactions. The name *travertine* comes from the Italian *travertino*, a corruption of *tiburtino*, meaning "stone of Tibur." Tibur was the original name of Tivoli, Italy, a famous travertine-producing locale. Travertine has long been used for pavement and building facings, and some of Italy's travertine quarries have been active for more than 2,000 years, supplying stone for such impressive structures as the Roman Coliseum

GETTING THERE

The Mammoth area contains some of the most impressive and active thermal features in Yellowstone. To get to stop 1 from the Mammoth Hot Springs Hotel, drive south for 2 miles (3.2 km) on Grand Loop Road and turn right (west) onto Upper Terrace Drive (32.2 miles, or 52 km, north of Madison Junction). After the turn, continue for about 0.2 mile (0.3 km) to the parking area on both sides of the road. From the parking lot, walk northeast on the boardwalk to New Blue Spring. Once you've had a chance to examine this hot spring, backtrack to the parking lot and proceed down the boardwalk to the southeast, toward Cupid and Canary springs. Along this stretch of boardwalk you'll see spectacular examples of travertine terraces and a hot springs pool associated with Canary Spring. Ancient examples of travertine similar to that forming at Mammoth today can be viewed up close in a landslide deposit at stop 2. This small roadside pullout is 1.7 miles (2.7 km) south of the turnoff to Upper Terrace Drive on the right (west) side of Grand Loop Road.

and the colonnades of St. Peter's Basilica. The light-colored rock consists of the minerals aragonite and calcite. Both minerals are made of calcium carbonate, although the arrangement of atoms within each is different. The microscopic crystals of aragonite and calcite in travertine grow in a variety of forms and patterns.

We'll start our tour at New Blue Spring. From stop 1, follow the sign and walk northeast down the boardwalk to the spring. One of the first things you will notice is the distinct odor of rotten eggs. This

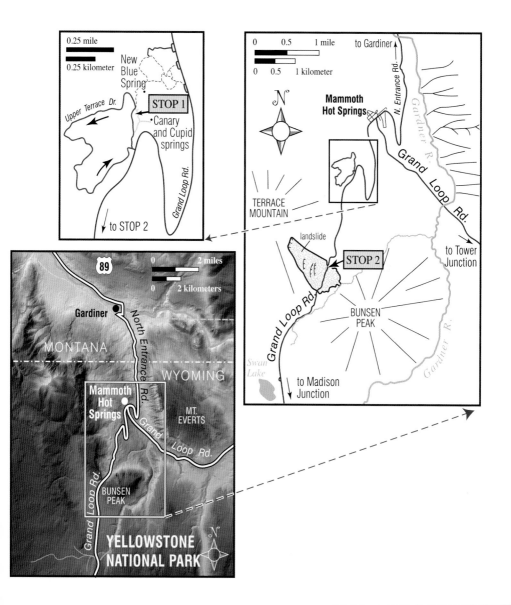

Travertine terraces at Mammoth Hot Springs viewed from the boardwalk near Cupid Spring. Many of the terraces are covered by small, scallop-shaped features called *shrubs*. Bacteria living in the thermal water produce the orange color.

odor is produced by sulfur and indicates that hydrogen sulfide gas is dissolved in the thermal waters. Many scientists have concluded that certain photosynthetic bacteria living in the hot springs use light to break down the hydrogen sulfide in the water, precipitating calcium carbonate in the process.

In addition, calcium carbonate precipitates directly from thermal water when its temperature and chemistry change. As the thermal water flows upward toward the surface and exits a vent, it experiences a significant drop in pressure. This causes carbon dioxide gas dissolved in the water to come out of solution in the same way it is released from a freshly poured beer—think of the foaming head of beer in a glass and the bubbles rising toward it, versus beer in a sealed bottle. The release of gas increases the pH of the thermal water, causing it to become supersaturated with calcium carbonate, which precipitates as the water flows downslope from the vent. This inorganic process occurs faster than that related to bacteria, leading some scientists studying the hot springs to suggest that most of the travertine at Mammoth is formed by inorganic precipitation.

Where does the carbon dioxide gas come from, and why is the thermal water at Mammoth saturated with calcium carbonate? These constituents of the thermal water are intimately linked, but let's start with the origin of the carbon dioxide.

In 2004, using evidence derived from the speed of earthquake waves passing through the bedrock beneath Yellowstone, scientists detected a big reservoir of carbon dioxide about 1.5 miles (2.5 km) beneath the surface of the northwestern part of the Yellowstone Caldera. The scientists inferred that the gas is coming from underlying magma associated with the Yellowstone Volcano. As the magma slowly moves upward into regions of lower pressure, carbon dioxide gas dissolved in it comes out of solution, partly filling pore spaces within the surrounding bedrock. Groundwater passing through the reservoir redissolves some of the carbon dioxide, in the same way that carbon dioxide gas is added to water and syrup to make soda pop.

The addition of the gas lowers the pH of the water, making it more acidic. Take a good swig of any carbonated beverage—soda, beer, what have you—and don't swallow for ten or twenty seconds. It hurts! The pain you feel is the acidity. As the acidic thermal water makes its way to the surface, it passes through a thick section of Madison limestone (see vignette 3) as well as older Cambrian limestone (see vignette 2). Although these limestones don't crop out much in the region around Mammoth, they are exposed in the Gallatin Range, and a small outcrop of Madison limestone also occurs near Roaring Mountain. These outcrops are like the tips of a large iceberg of limestone beneath the surface of much of northern Yellowstone National Park. The hot, acidic water dissolves some of the calcium carbonate composing the limestone and carries it along in solution. By the time the thermal water flows out of the vents at Mammoth, it is saturated with carbon dioxide gas and calcium carbonate.

Most travertine deposits are associated with faults that provide pathways for mineral-rich fluids to reach the surface. Geologists have mapped numerous faults in the Mammoth area, and the largest are interpreted to reach more than 5 miles (8 km) into the subsurface.

Based on similarities in the types and concentrations of gases dissolved in the water at Mammoth and Norris Geyser Basin, located about 16 miles (25.7 km) to the south, some geologists have suggested that one set of faults provides a pathway for thermal water to travel from Norris to Mammoth, where it might account for up to 40 percent of the fluid and heat discharging there. Others don't accept the fault connection, instead maintaining that a local body of magma about 2 miles (3.2 km) beneath the surface at Mammoth or the deep circulation of thermal water to depths exceeding 6 miles (10 km) is sufficient to explain the composition of the thermal water and the amount of heat being discharged at Mammoth.

Regardless of what's going on beneath the surface, the travertine features at the surface are a wonder to behold—and varied. Different crystal structures form in close proximity. For example, the crystals that form the rims of the terraces are different from those forming in the pools only a few steps away. Combined with frequent shifts in the location of individual terraces, these small-scale variations make travertine an unusual-looking rock and provide much of its appeal as an ornamental stone. Viewed up close and in cross section, ornamental travertine typically is a layered rock in which the layers are less than 1 inch (2.5 cm) thick. Some layers are expressed as simple, relatively continuous lines, whereas others are discontinuous and contain an abundance of small holes or other intriguing features. The formation of travertine and travertine terraces is a complicated process involving changes in water chemistry, the presence of bacteria, and the temperature and degree of agitation of the thermal water.

With that in mind, let's look at New Blue Spring, one of the main vents from which thermal water often escapes at Mammoth. Although thermal water is escaping mainly from the uppermost terrace, vents frequently change locations as the precipitation of calcium carbonate seals them, forcing the water to find a new opening. Each terrace around the spring has a lip running along its upper edge, which impounds water behind it. Typically, the pools that cover the tops of the terraces are only 1 foot (30 cm) or less deep. Where water spills over a terrace and flows down the vertical wall below it, bizarre, icicle-like forms of white or orange travertine often develop. Numerous 1-inch-long (2.5 cm) scalloped bumps project from the walls of others.

Unless New Blue Spring is dry, it's not possible to look directly into the pools there, although you can still see that they do not have smooth bottoms, but rather are covered by numerous small bumps that are more circular in shape and smaller than those that occur on the vertical walls of the terraces. Both forms are called *shrubs*. Composed of calcium carbonate crystals, the shrubs forming in the pools typically are less than 1 inch (2.5 cm) wide and about as tall. On occasion, usually late in the summer, New Blue Spring will cease to flow and the pools near the vent will dry out, providing a good opportunity to see the shrubs.

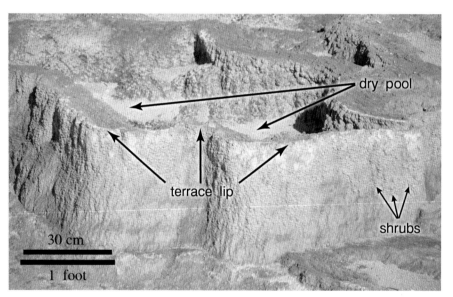

A dry pool near New Blue Spring in late summer.

Detailed analysis of shrubs from Mammoth and elsewhere has shown that they occur in a wide variety of shapes and patterns. At one extreme are bacterial shrubs, which look something like bushes with poorly defined branches. When viewed with a scanning electron microscope, bacterial shrubs are shown to be loaded with tiny holes that geologists interpret to be molds of rod-shaped bacteria, which are thought to significantly influence the growth of the shrubs. Crystal shrubs, however, have well-defined branches and few of the tiny holes interpreted to be bacterial molds. Geologists infer that the growth of

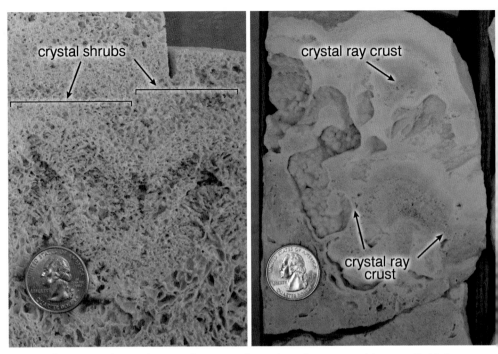

Cross sections of travertine samples taken from near Mammoth Hot Springs. The sample on the left shows two small calcium carbonate shrubs. The best-developed shrub is directly above the quarter. The sample on the right shows crystal ray crusts that probably developed near a hot spring vent. The bumps inside the open pores on the left side of the sample are smaller crystal ray crusts. Quarters for scale.

The view east toward Canary Spring. The water flows from the spring downslope to the Gardner River.

these shrubs is more strongly influenced by the inorganic precipitation of calcium carbonate. Bacterial shrubs form in the relatively quiet water of the terrace pools, whereas crystal shrubs form on the vertical walls as water spilling over the walls rapidly releases its dissolved carbon dioxide gas.

The rims of travertine that run along the outer edge of each terrace and form the lip of a terrace pool arise from the especially rapid precipitation of calcium carbonate. In the same way that the vigorous stirring of a freshly opened beer releases small bubbles of carbon dioxide gas, thermal water releases carbon dioxide as it flows away from the vent and is agitated. The release of carbon dioxide causes the thermal water to become supersaturated with calcium carbonate, which it precipitates, typically in an inorganic form called *crystal ray crust*. Using computer modeling, some scientists have concluded that the angle of the slope over which thermal water flows after leaving a vent determines whether or not a terrace develops. If the slope is steep enough, the water flows turbulently downward, releasing carbon dioxide and forming crystal ray crust that grows upward and outward at rates exceeding 1 inch (2.5 cm) per year, faster than the upward growth of the pool floors behind them.

Backtrack to the parking lot and walk southeast down the boardwalk toward Cupid and Canary springs, following the signs. Depending on the amount of water being discharged from Mammoth's springs, this

runoff channel with crystal shrubs and small terraces

stretch of boardwalk passes close by one or more small streams that eventually empty into the Gardner River. The streams are usually less than 1 inch (2.5 cm) deep but can be 10 feet (3 m) wide, and most have numerous small terraces generally less than 1 inch (2.5 cm) high. As with the larger ones, the smaller terraces are produced as water is agitated, causing calcium carbonate to build up more in the zones of turbulent water. As the rim of travertine builds upward and outward, water flowing around its margins extends the rim laterally.

As you approach Canary Spring, the boardwalk passes by a pond of thermal water on the left that is partially impounded by a rim of travertine. If you look into the pond, you'll notice that sticks, pine cones, pieces of grass, and other organic debris at the bottom of the pond all are being actively encrusted within a white rind of calcium carbonate. The calcium carbonate precipitates faster than the organic material decays, so the debris is incorporated into the travertine deposit. Eventually, the organic objects decay, leaving behind molds.

Thin flakes of white calcium carbonate resembling broken potato chips also occur on the pond bottom. Most are 1 to 2 inches (2.5 to 5 cm) across and about $\frac{1}{16}$ inch (1.6 mm) thick. Many are not attached to the bottom. The flakes formed from terraces upslope that dried

Many of the ways calcium carbonate precipitates in thermal water are visible in the pond just above Canary Spring. Sticks are being encrusted by calcium carbonate, and a variety of shrubs—the dime- to quarter-sized bumps that look like heads of cauliflower—are growing on the floor of the pond. Miniature terraces develop where thermal water is flowing over the bottom of the pool.

out and flaked apart. Water in the streams picked up the flakes and deposited them here. Numerous dime- to quarter-sized bumps up to 0.5 inch (1.3 cm) high and resembling miniature heads of cauliflower also are visible on the bottom. These are calcium carbonate shrubs. Some have tiny offshoots, or branches, that typically are less than ⅛ inch (3 mm) long.

You'll also probably see a very thin white scum of calcium carbonate floating on parts of the pond. Some of these patches contain small mineralized bubbles that originally held pure oxygen released by photosynthesizing algae that live in hot springs. Thermal water enveloped the bubbles and coated them with calcium carbonate.

Only small shrubs occur in those parts of the pond the boardwalk passes over, so it can be difficult to appreciate them. Fortunately, good views of similar shrubs can be seen at stop 2, where large blocks of ancient travertine can be examined in a landslide deposit. Many of the blocks are broken perpendicular to the layering in the travertine, exposing cross-sectional views of features that can be seen only from the top down in the modern travertine deposits at Mammoth. You may want to take a few minutes to wander around the field of angular travertine boulders at stop 2, studying some of the broken surfaces.

Radiometric dating of travertine exposed upslope at Terrace Mountain, the source of the landslide at stop 2, indicates that the rock formed between 390,000 and 360,000 years ago, making it the oldest travertine in the Mammoth-Gardiner area. Nearly all other travertine in this region is 57,000 years old or younger, including the terraces at Mammoth, which are generally less than about 7,000 years old.

Perhaps the first thing you'll notice is that the travertine is clearly layered, and that the layering accentuates tear-pants weathering on many of the rock surfaces (see vignette 3 for more on "tear-pants weathering"). The boulders you want to examine in detail are those that are freshly broken and display a less-weathered surface.

The freshest surfaces show clearly defined layering in which individual layers typically are less than 1 inch (2.5 cm) thick. Small holes that are elongate and parallel to the layering are common. Think back to the pond above Canary Spring. These pores are created when organic debris (twigs, algae, pine needles, and so on), potato chip–like fragments of calcium carbonate, and bubbles are incorporated into a travertine

Many of the holes in this travertine probably existed when the rock first formed but have been enlarged by the weathering process. Most of the boulders at stop 2 display tear-pants weathering like this. Penny for scale.

deposit. Many of the pores are coated with very tiny crystals that grew inward from the walls of the pores. These crystals indicate that calcium carbonate continued to precipitate after the pores had formed.

A bizarre, honeycomb-like feature is probably the most striking aspect of the ancient travertine. Look for clusters of pores within one layer that are stretched out perpendicular to the layering. Each pore has the same approximate dimension, typically about ⅛ inch (3 mm) across and 1 inch (2.5 cm) long, with rounded ends. Geologists interpret these features to be clusters of mineralized bubbles that formed on the floor of a hot spring. Calcium carbonate precipitated around the bubbles as they clung to the bottom of the hot spring. As the floor of the hot spring built upward due to the precipitation of more calcium carbonate, the bubbles grew upward at the same rate, leaving behind the elongate bubble holes.

Some of the boulders at stop 2 display good examples of shrubs. You have to look closely to see them; the shrubs usually are less than 1 inch (2.5 cm) high and contain numerous branches. Some layers within the rock consist entirely of shrubs. The presence of multiple shrub layers reflects the upward growth of a terrace pool. Many of the shrub layers are separated by thinner layers of less-porous calcium carbonate that must have formed under different conditions than were present when the shrubs grew. Some geologists have suggested that these differences reflect seasonal changes in the temperature and chemistry of a hot spring, which in turn affected the type of calcium carbonate structure that formed.

The travertine deposits in the Mammoth area are unique to Yellowstone Country, mainly because the limestone that supplies the calcium carbonate needed to form travertine is relatively close to the surface and there are deep faults along which mineral-rich water can travel

Geologists interpret these honeycomb-like features to be clusters of mineralized bubbles. Two layers containing bubble clusters are visible. Penny for scale.

to interact with the limestone. In contrast, most of the other famous thermal features in Yellowstone National Park, including its geysers, form from a different type of rock called *sinter*. In vignette 18, we turn our attention to extensive sinter deposits in Upper Geyser Basin, one of the most popular thermal areas in Yellowstone.

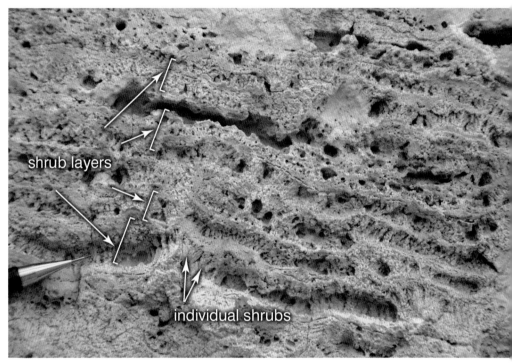

shrub layers

individual shrubs

Several layers of shrubs exposed in travertine at stop 2. Pencil tip for scale.

18.
Siliceous Sinter
Old Faithful and Upper Geyser Basin

According to the U.S. Geological Survey, Yellowstone National Park contains about two-thirds of Earth's geysers, thermal features from which hot water and steam periodically erupt. None of Yellowstone's geysers are more famous than Old Faithful Geyser, named in 1870 by members of the Washburn Expedition. A brief walking tour in Upper Geyser Basin provides a good opportunity to see Old Faithful, one of the park's crown jewels, as well as some of the different types of thermal features Yellowstone contains.

We begin at Old Faithful Geyser, whose name and fame both reflect the fact that the geyser erupts with some regularity, although not as much as you might think. Not only is there about twenty minutes of uncertainty in the estimated timing of each eruption, but long-term records indicate the average length of time between eruptions has increased. The National Park Service kept records of Old Faithful's activities between 1937 and 1956. Park rangers observed most of its eruptions during this period, and their records show that the average interval of time between eruptions ranged from a high of just over sixty-seven minutes in 1937 to a low of sixty-two minutes in 1949. Record-keeping resumed in 1997, and by that time the average interval had increased to seventy-five minutes. The interval continued to increase until 2002, when it leveled off at just over ninety minutes. As of 2010, eruptions occur, on average, ninety-one minutes apart.

Even though the average amount of time between its eruptions has lengthened, the eruptions from Old Faithful Geyser are still widely regarded as being the most predictable and regular in the park (more

GETTING THERE

Old Faithful Geyser, stop 1, is one of many thermal features in Upper Geyser Basin. To get there, proceed to Old Faithful, located on Grand Loop Road 17.6 miles (28.3 km) northwest of West Thumb and 15.9 miles (25.6 km) south of Madison Junction. Park in one of the lots near the visitor center and follow the signs to the geyser. After viewing Old Faithful, walk north on the boardwalk to where it meets a paved walkway. Follow the pavement to where it forks and take the right fork, which leads to a bridge over the Firehole River. Cross the river and follow the boardwalk for about 400 yards (360 m) to Pump Geyser, stop 2. The strikingly blue Doublet Pool, stop 3, is located about 100 yards (90 m) farther northwest along the same stretch of boardwalk.

An eruption of Old Faithful Geyser, which draws tourists from all over the world. During busy summer months, the National Park Service posts a sign in the Old Faithful Visitor Center that estimates the time of the next eruption.

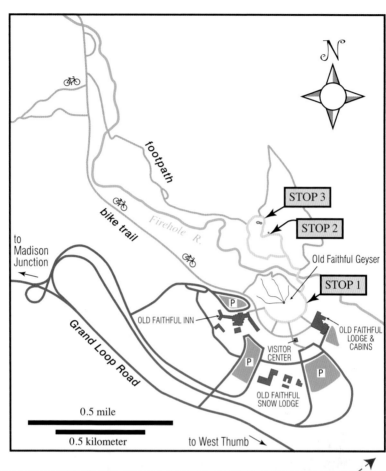

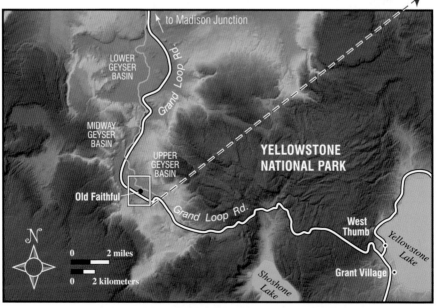

on this later). But what causes the geyser to erupt in the first place? To answer this question, it is necessary to think about the internal structure of the geyser and, more generally, how geysers work.

Unless you happen to arrive during an eruption, Old Faithful Geyser is a cone-shaped low hill of light-colored rock with steam emanating from a hole, called a *vent*, at its summit. A conduit connects the vent to a network of openings that project into the subsurface, forming a plumbing system that can extend downward hundreds or thousands of feet (hundreds of meters) and provides a direct connection between the surface and hotter rocks at depth. In the Yellowstone region, a body of magma as little as 2 miles (3.2 km) beneath the surface provides the heat (see vignette 11).

Geysers erupt because of the interaction between heated groundwater and the shape of the conduit. After a geyser erupts, water begins refilling the plumbing system. Some is erupted water that drains back into the vent, some is shallow groundwater, and some is thermal water circulating upward from greater depths. As water fills the plumbing system, pressure builds. This is the same type of pressure you feel at the bottom of a deep swimming pool with all that water above your head.

Water in the deepest parts of a geyser's plumbing system is heated well beyond its normal boiling point. This superheated water doesn't convert to steam and boil like water in a heated kettle because the weight of cooler water overhead exerts a confining pressure. As superheated water circulates toward the surface where there is less pressure, however, some of it does turn to steam. The rising steam and superheated water heat the cooler water that resides higher in the plumbing system, producing more steam. Some of this steam escapes out the geyser's vent, explaining why Old Faithful releases steam even when it is not erupting.

Not all of the steam that is generated can escape out the vent. Some becomes trapped at one or more constrictions in the plumbing system, forming bubbles of steam that grow larger. Eventually, the growing steam bubbles force water out of the vent, lowering the pressure in the conduit below and causing a large amount of superheated water to rapidly boil. The rapid boiling produces the full-on eruption, in which liquid and steam are forced out of the geyser's vent like water from a garden hose. As water is ejected, pressure in deeper parts of the plumbing system is released, causing water there also to flash to steam. This process reaches into the deepest parts of a geyser's plumbing system until all the water is gone or the temperature drops sufficiently to prevent further mass boiling.

There is a balance between several factors that must occur if a geyser is to form. If too little water fills the plumbing system, a geyser will not erupt but instead will become a fumarole—a plume of steam escaping from a vent. If too much water is available or the heat source is insufficient to superheat enough water, the result will be a hot spring. If the plumbing system does not include any major constrictions to delay the escape of steam to the surface, not enough pressure will build to initiate an eruption. In the case of Old Faithful, these factors have combined to produce a geyser that erupts with predictable regularity.

Between 1983 and 1994, geoscientists undertook an intensive study of Old Faithful. They lowered a video camera, thermometers, and pressure gauges into the uppermost portions of the geyser's conduit. The thermometers and pressure gauges reached depths of nearly 70 feet (21.5 m), while the video camera went as far as 46 feet (14 m) below the vent. They found that Old Faithful's conduit includes cracks, small caverns, and twisting passageways but is dominated by a single east-west-oriented, mostly vertical opening. About 21 feet (6.4 m) below the surface geologists found a narrow place where the conduit is only about 4 $\frac{1}{8}$ inches (10.5 cm) wide and no more than 6.5 feet (2 m) long. They interpreted this as a choke point, a significant constriction in the conduit that causes Old Faithful Geyser to erupt high into the air. Scientists estimate that during the peak of an eruption fluids are traveling past the choke point at about 230 feet per second (70 km per second), or 157 miles per hour (253 km per hour).

Below the slotlike choke point the conduit opens up into a small cavern about 2 feet (60 cm) wide, with a waterfall spilling into it from the side. The scientists observed two small ledges below this cavern. Below the second ledge was a larger cavern more than 3 feet (0.9 m) across and more than 10 feet (3 m) deep. A slot less than 1 foot (30 cm) across marking the base of the cavern was the deepest feature observed with the camera. Analyzed together, the information from the video camera, thermometers, and pressure gauges revealed much about the shape of Old Faithful's conduit, how it refills between eruptions, and the sequence of events that lead to an eruption.

Viewed from the surface, Old Faithful's eruptions typically follow the same basic script. The beginning of an eruption is signaled by a few small spurts of water leaving the vent, forced upward by growing bubbles of steam trapped in constrictions below. The squirts quickly grow in size as the small release of pressure caused by the escaping water allows superheated water below to flash to steam. The peak of the eruption is marked by a vertical jet of water and steam that reaches heights between 106 and 184 feet (32 and 56 m) and usually lasts about 20 to 30 seconds. This part of the eruption releases the water and steam that were trapped below the choke point. Water falls back to the

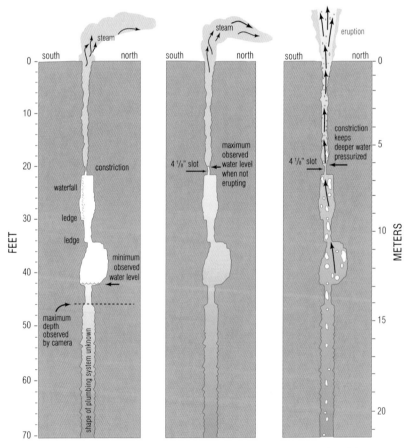

Cross section of Old Faithful's conduit, as determined by researchers who lowered a camera into the main vent several times between 1983 and 1994.

ground near the vent, landing with an audible splash. As the eruption wanes, the height of the water plume lessens before degenerating into a few last spurts. Eventually, only steam remains. Over the course of an entire eruption, which is 1.5 to 5 minutes long, the geyser spews out between 3,700 and 8,400 gallons (14,000 and 31,800 liters) of water. The upper end of this range is enough to fill a pool 5 feet deep, 10 feet across, and a little more than 22 feet long (1.5 by 3 by 6.7 m).

In addition to variations in the amount of water escaping Old Faithful Geyser during an eruption, there are also variations in the average interval of time between eruptions. Geoscientists interpret longer-term changes in this interval, which occur over several years or decades, to reflect alterations in the pressure that builds in the plumbing system.

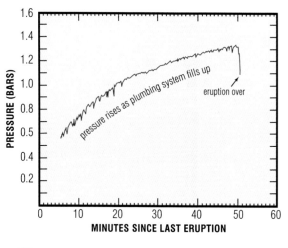

Graph of pressure buildup and release recorded during one eruption of Old Faithful Geyser. The large pressure drop represents the eruption itself, and the small dips prior to the eruption reflect releases of steam bubbles that had built up as the geyser was refilling.

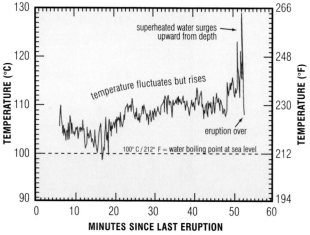

Graph of water temperature during one eruption of Old Faithful as measured within the conduit about 69 feet (21 m) below the surface. The fluctuations reflect the mixing of the relatively cool water near the surface with hot water from deeper in the plumbing system as the geyser was refilling. The very hottest temperatures were recorded during the eruption as superheated water rushed upward past the thermometer.

These pressure changes depend, in part, on the amount of surface water that is available to refill the plumbing system. Scientists have recognized a link between the frequency of eruptions of Old Faithful and the average annual discharge of the Madison River. The flow of the Madison River is directly related to the amount of annual precipitation in its headwaters, which includes Upper Geyser Basin. During years when the Madison River flow is high, eruptions of Old Faithful Geyser are more frequent because more water is available to refill the plumbing system, leading to more rapid pressure increases. During low-flow years, less water is available, leading to slower pressure increases and less frequent eruptions.

Other variations in Old Faithful's eruptive interval are shorter term, such as monthly variations that occur with the changing seasons. The longest intervals typically occur in the spring and summer, during and after peak runoff in nearby rivers. This relationship suggests that each year a fresh slug of relatively cool water infiltrates Old Faithful's plumbing system, cooling its water and, as a result, increasing the interval between eruptions. Less cool water infiltrates the system during the fall and winter, allowing water in the geyser's plumbing system to heat more quickly and the geyser to erupt more frequently.

Earthquakes can also affect the eruptive behavior of Yellowstone's geysers. Old Faithful's eruptive interval increased immediately after the 1959 Hebgen Lake earthquake (discussed in vignette 20), and again after the 1983 Borah Peak earthquake centered in Idaho. Although it did not affect Old Faithful, a strong earthquake in 2002 near Denali National Park in Alaska appears to have altered the eruptive behavior of eight of Yellowstone's geysers, including Daisy Geyser in Upper Geyser Basin.

Unlike the changes in pressure brought on by variations in the amount of available water, earthquakes physically alter the plumbing systems of Old Faithful and other geysers. Particularly important are the relatively slow surface waves that earthquakes generate. In contrast to body waves, which travel through the Earth, surface waves travel along the ground surface, causing it to move slowly up and down and to and fro, much like swells in the ocean. The ground surface and shallow subsurface are alternately squeezed and pulled apart, and sheared one way and then the other. Although the amount of deformation can be so small that sensitive instruments are required to detect it (the

Denali earthquake was not felt by people in Yellowstone but it was registered on seismic instruments there), surface waves can loosen and dislodge sediment and rock within the plumbing systems of geysers. This affects the ability of water to move around, which can alter a geyser's eruptive behavior.

The longest-term changes in geyser eruptive behavior are caused by the deposition of minerals on the walls of the plumbing system. Unlike the hot springs deposits at Mammoth, which are made of travertine (see vignette 17), the light-colored rock of the thermal features of Upper Geyser Basin is made of silica. Called *siliceous sinter*, it is a porous chemical sediment deposited by springwater. Nearly all of the silica in the sinter at Upper Geyser Basin is amorphous—that is, the silica molecules are not arranged in an orderly atomic fashion and do not form crystals, although they can form small crystal-like structures called *spicules*.

Why are the thermal features at Mammoth made of travertine, while those in Upper Geyser Basin are made of sinter? The answer lies in the type of rock beneath the two thermal areas. The rocks below Mammoth include thick sequences of limestone. In contrast, the rock underneath Upper Geyser Basin is rhyolite that filled the caldera of the Yellowstone Volcano (see vignette 13). As superheated water rises from deep below the surface, it passes through the rhyolite, dissolving silica in the rock and becoming supersaturated with it. As the water nears or reaches the surface, it cools and partially evaporates, precipitating silica in the form of siliceous sinter. That's how the mound of sinter surrounding the vent of Old Faithful was produced: Silica-saturated water erupting out of the geyser splashed back to Earth, cooled, and partially evaporated, depositing silica. Because most of Old Faithful's water lands close to its vent, the sinter has built up into a cone-shaped hill.

The video recordings of Old Faithful's conduit showed dime- to quarter-sized bumps of silica growing on its walls. Although scientists were not able to sample these, silica collected from other geysers shows that it precipitates in concentric layers around the inner walls of a conduit. Each layer is made of densely packed spicules that grow parallel to each other and perpendicular to the layering. Viewed in cross section, the spicules growing together resemble a palisade-type fence.

The tightly packed character of the sinter, and the fact that the spicules grow perpendicular to the walls of the conduit, help hold it in place during violent eruptions. As the spicules grow toward the center of a conduit, they alter its shape, creating constrictions and choke points and ultimately sealing the conduit, forcing superheated water to abandon the vent and find a different path to the surface.

There are at least two other cone-shaped sinter mounds near Old Faithful, and neither has steam escaping its summit. The mounds mark former geyser vents that were abandoned, probably because the conduit below each vent became plugged with precipitating silica. It's likely that these mounds were ancestral versions of Old Faithful, and that as one conduit became plugged, the pressurized water simply found a nearby route to the surface, giving birth to the next geyser. The mounds suggest that this part of Upper Geyser Basin has a long history of geyser activity. Although no one knows how long Old Faithful Geyser will remain active, eventually its vent will become plugged with sinter. Perhaps another geyser will form nearby, or maybe this part of Upper Geyser Basin will no longer have a geyser. If a new geyser does form, differences in the shape of its conduit almost certainly will produce a geyser with a different eruptive behavior from Old Faithful Geyser.

The plume of steam in the background is Old Faithful Geyser, seen here between eruptions. The inactive sinter mound in the foreground, one of several near Old Faithful, suggests that one or more ancestral versions of Old Faithful Geyser may have existed, with the position of the geyser shifting as the conduit beneath each mound became plugged by sinter.

Old Faithful is a classic example of a cone geyser, characterized by nearly vertical jets of water and steam, a prominent cone, and little or no standing water around the vent. Pump Geyser at stop 2 is a great example of a fountain geyser, characterized by eruptive bursts of water and a shallow moat of standing thermal water surrounding the vent.

Among the first things you will notice about Pump Geyser is that the water in it is constantly surging in a manner reminiscent of a mechanical pump. The pumping action is related to how quickly the geyser's conduit refills—typically within a few seconds. Each surge is an eruption caused by the escape of hot water and steam from the conduit. Unlike Old Faithful, Pump Geyser's eruptions are broader, and not all of the water shoots straight up. Much of the water flows back into the vent following each surge, but some ends up in the surrounding moat or runs off in a small stream that passes under the boardwalk.

The forms of the sinter around the geyser's vent differ depending on their proximity to the vent and the small stream of runoff produced by the geyser. On the far side of the vent are smooth, yellow-colored deposits that are similar to the bumps of silica recorded in the conduit of Old Faithful Geyser. The bumps are either completely submerged by water in the vent or are constantly wet due to the surging action of the geyser and the drifts of steam that pass over them. Similar bumps sampled from other fountain geysers were made of elongate spicules that grew perpendicular to the surface of the bumps and formed the same palisades-type structure observed in the interior of cone geysers.

More bizarre are the small knobs that occur on the nearest rim of the geyser's vent and, to a lesser degree, in and along the small stream of water running from the geyser. Some resemble mushrooms. Unlike the bumps that are constantly wet, the knobs are episodically wet and dry. When water is splashed on them it evaporates and deposits a thin layer of silica. Silica precipitates more quickly in places where wetting and drying occurs than it does in places that are constantly wet; as a result, silica precipitates more quickly on the knobs than on the regions between them that don't dry out.

In addition to a wide assortment of cone and fountain geysers, Upper Geyser Basin also contains numerous hot springs, such as Doublet Pool, at stop 3. Perhaps the most striking attribute of the pool is its deep blue color, a characteristic of many of the hot springs in the

Pump Geyser erupts once every few seconds. The rim of yellow rock surrounding the geyser's vent is siliceous sinter, as is the white rock farther away. The larger, quarter-sized bumps on the far side of the vent are typical of sinter that forms in wet conditions. The slightly smaller, knobby, mushroomlike features on the near side form in conditions that alternate between wet and dry *(inset)*.

basin. When light interacts with water, it causes a complex series of vibrations within the water molecules. Each type of vibration absorbs the energy of certain wavelengths of light. The absorption of red wavelengths gives water its characteristic blue color.

However, the blue of many of Yellowstone's thermal waters, including Doublet Pool, is considerably more intense than the water in lakes, oceans, and swimming pools because of the dissolved silica it contains, which causes additional scattering of the blue part of the visible light spectrum. Recently published work has shown that longer chains of silica molecules dissolved in water are very effective at scattering blue wavelengths of light. *Long* is relative, considering that these chains are between 4 millionths and 200 thousandths of an inch (between 0.1 and 0.5 micrometers) in length. Because the blue light entering the pool is scattered, more of it reaches your eye, giving the water a deep blue color.

The water below the surface in the deepest parts of Doublet Pool also appears to shimmer faintly. It is very similar to the shimmering that is produced when sugar water or saltwater is poured slowly into a glass of freshwater. The density of the two fluids differs, affecting the angle at which light passes through the mixing fluid and causing it to shimmer. Likewise, local variations in the amount of dissolved silica in Doublet Pool cause variations in the density of its water. Eddies and swirls in the slowly mixing fluid cause light passing through it to bend at different angles, producing the shimmering.

Dissolved silica scatters blue light waves and is responsible for the striking blue color of Doublet Pool. The silica also precipitates as the white sinter crust around the margins of the pool.

The high concentration of silica here also results in the growth of the curious white crust you can see surrounding the pool. The crust is characterized by numerous lobe-shaped protrusions that extend from the edges of the pool out and over the water surface up to about 4 inches (10 cm). The lower surface of the crust is approximately level with the water surface and is kept constantly wet, whereas the top of the crust is 1 inch (2.5 cm) or so higher, slopes slightly away from the edge of the pool, and is usually dry. The tips of the lobes are bright white, whereas the crust about 4 inches (10 cm) away from the water's edge is grayish because the sinter there is weathering. Within about 1 foot (30 cm) of the water's edge, the crust is covered by loose sediment formed from weathering of the surrounding ground surface.

Researchers from Stanford University studied similar crusts in hot springs in Lower Geyser Basin and found that the crust is a very porous, laminated sinter. They also found that seasonal changes in air temperature alter the rate at which sinter precipitates as crust. More sinter precipitated during the fall, winter, and spring, whereas little growth occurred during the summer. Although sinter growth is still being studied, other researchers have shown that concentrations of dissolved silica in thermal water decrease with temperature, suggesting that silica precipitates more readily in cooler water.

The fringing crust at Doublet Pool and the impressive assortment of thermal features in Upper Geyser Basin all are the result of actively forming sinter, which is slowly changing the internal and external shape of the features. In addition to the relatively slow physical changes brought about by precipitating sinter, some of Yellowstone's thermal features have experienced very rapid changes in the form of hydrothermal explosions—the topic of vignette 19.

19.
Hydrothermal Explosions
Norris Geyser Basin and Yellowstone Lake's North Shore

Each year millions of tourists are drawn to Yellowstone National Park to see its world-class collection of thermal features. Among the geysers, travertine terraces, hot springs, and mud pots that draw most visitors are lesser-known features—craters in this case—that formed during hydrothermal explosions, violent ejections of water, steam, rock, and mud from the ground.

About twenty-five geologists and other scientists on a field trip witnessed this hydrothermal explosion in Biscuit Basin in 2009. The explosion took place without warning and sent debris about 50 feet (15 m) into the air.
—Photo courtesy of Wade Johnson

Hydrothermal explosions are not well understood, in part because they are relatively rare events. A 2001 worldwide compilation of craters created by hydrothermal explosions identified only thirty-one historic and forty-seven prehistoric craters outside of Yellowstone National Park; all have diameters less than 550 yards (500 m), and all but nine have diameters less than 110 yards (100 m). In contrast, the U.S. Geological Survey and National Park Service have identified and described more than twenty large craters within Yellowstone, including ten with diameters greater than 550 yards (500 m) and one with a length of 1.7 miles (2.7 km) and width of 1.5 miles (2.4 km). They also identified dozens of small craters measuring less than 11 yards (10 m) across that probably were caused by hydrothermal explosions. Park visitors have witnessed several relatively small hydrothermal explosions, most recently in Biscuit Basin in 2009. Small explosions like these occur about once every other year, whereas larger explosions have occurred, on average, once every 700 years.

Hydrothermal explosions occur in areas that contain substantial subsurface reservoirs of superheated water—water that is above the

---

GETTING THERE

Norris Geyser Basin is the most active thermal area in the Yellowstone region. Located off Grand Loop Road, the basin is 21 miles (33.8 km) south of the junction of Grand Loop Road and North Entrance Road in Mammoth Hot Springs and 13.3 miles (21.4 km) northeast of Madison Junction. Park in the main parking lot and continue on foot past the visitor center and to the left, heading southwest into Back Basin. Stop 1 is Pearl Geyser, located along the western part of the Back Basin boardwalk. About 150 yards (135 m) southwest of Pearl Geyser is stop 2, Porkchop Geyser, which is surrounded by large, angular boulders. Stop 3 is Indian Pond, a hydrothermal explosion crater close to the north shore of Yellowstone Lake. To get there from Norris, drive 11.6 miles (18.7 km) east on Norris Canyon Road to Canyon Village. Turn right (south) on Grand Loop Road and proceed 15.4 miles (24.8 km). Turn left (east) on East Entrance road and proceed about 3.2 miles (5.1 km) to one of the small parking lots on the right (south) side of the road. To reach the hydrothermal explosion breccia deposit, from the parking lots follow the footpath southeast, past Indian Pond, to the north shore of Yellowstone Lake. Scramble down one of the gullies to the shoreline. Stop 4 is Steamboat Point, which is 3.2 miles (5.1 km) farther east along East Entrance Road.

---

boiling point but does not boil because the weight of the water above it exerts a confining pressure. If the confining pressure is reduced, however, superheated water can rapidly flash to steam. Unlike an erupting geyser, in which steam forces the water above it through a preexisting

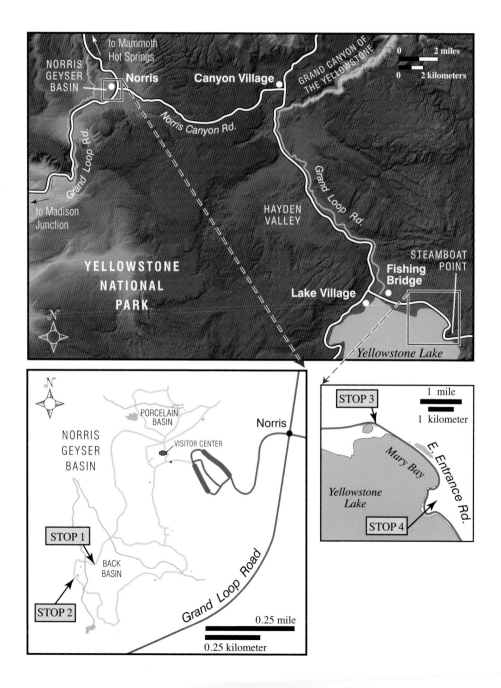

conduit and out a vent (see vignette 18), a hydrothermal explosion occurs when superheated water suddenly flashes to steam, fragmenting overlying rock and propelling it skyward along with mud, water, and steam. The initial removal of the rock mass reduces the confining pressure at deeper levels in the subsurface, allowing superheated water there to flash to steam. The result is a cascading series of explosions that can take place over minutes or hours, although many small hydrothermal explosions consist of a single blast. As the pressure release reaches deeper into a reservoir of superheated water, the steam that is generated is unable to expel the overburden that remains, and eventually the explosions stop. Studies of modern and ancient hydrothermal explosions and their aftermath suggest that the explosions can start at depths as shallow as about 15 feet (4.6 m) and reach as deeply as about 1,000 feet (300 m) below the surface.

Norris Geyser Basin is located just outside the northwestern edge of the Yellowstone Caldera in an area where numerous north-trending normal faults intersect the ring faults that define the caldera's margin. The normal faults dip steeply into the subsurface. Movement along the intersecting faults has significantly fractured the rocks in the subsurface, allowing thermal fluids to rise from deep below the surface and escape via the basin's thermal features.

Geochemists have suggested that Norris's thermal water comes from several interconnected sources that reside at different depths. They deduce that the deepest source is 0.5 mile (0.8 km) or more below the surface, clear blue, and more than 520 degrees Fahrenheit (270 degrees C), although some estimates suggest it exceeds 570 degrees Fahrenheit (300 degrees C). Situated above this hot water is a reservoir of somewhat cooler water, estimated to be less than 480 degrees Fahrenheit (250 degrees C). Relative to the water below it, this reservoir is thought to be acidic and contains suspended clay derived from volcanic rocks in the subsurface. Above this acidic water is cooler surface runoff that percolates into the subsurface through a network of interconnected pores and cracks. Surface runoff percolating downward first passes through pores and cracks mostly filled with air, although a little deeper the pores are filled with water. The boundary between the air-filled and water-filled pores is called the *water table* and is an important feature for understanding why hydrothermal explosions occur.

A

ground surface

hot springs

water table

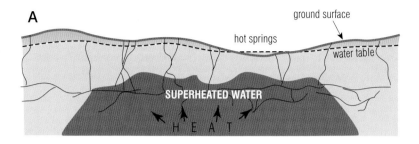

SUPERHEATED WATER

H E A T

B

ground surface

water table drops

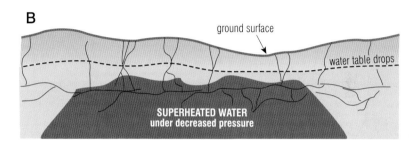

SUPERHEATED WATER
under decreased pressure

C

hydrothermal explosion

explosion front propagates downward

D

crater

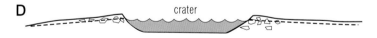

The evolution of a hydrothermal explosion. (A) Confining pressure
keeps superheated water from flashing to steam. (B) A drop in the
water table reduces confining pressure, causing relatively shallow
superheated water to flash to steam, which fragment the overlying
rock and propels it upward (C) along with mud, steam, and water.
Sudden removal of the overburden causes superheated water to
experience a drop in confining pressure, and it flashes to steam,
producing a cascading series of explosions that lasts until the steam
is no longer sufficient to remove the remaining overburden. (D) A
crater rimmed by ejected debris remains and often fills with water.

Norris Geyser Basin has long been regarded as the most geologically unstable of Yellowstone's thermal areas. Nearly every year, thermal features across the basin undergo multiple, simultaneous, and generally temporary changes during an event referred to as the *geothermal disturbance*. The disturbance has been the subject of considerable study. The first recorded observations of the disturbance were made in 1926. Subsequent observations have shown that nearly all instances occurred in August or September. Based on weekly sampling and monitoring of eight different thermal features in Norris Geyser Basin in 1995, geoscientists from the U.S. Geological Survey concluded that the geothermal disturbance reflects a seasonal lowering of the water table. This lowering decreases the confining pressure in the reservoir of acidic, muddy, superheated water, so some of it flashes to steam. Shortly thereafter, some of the water in the deeper reservoir flashes to steam as the pressure there is reduced.

Although insufficient to produce a hydrothermal explosion, the subterranean boiling during the disturbance mixes water from the two reservoirs and causes acidic, muddy water to discharge from thermal features that normally are dominated by the clear blue water of the deeper reservoir. The mixing also leads to a host of other changes in the basin's thermal features: temperatures fluctuate, some geysers erupt more frequently, and many hot springs boil more energetically while others dry up and become vents of hissing steam called *fumaroles*.

With the geothermal disturbance in mind, let's start our tour at stop 1, located at Pearl Geyser in a part of Norris Geyser Basin called Back Basin. This relatively small fountain geyser (see vignette 18), marked by a pool of water that is about 20 feet (6 m) across and at least 3 feet (0.9 m) deep, has undergone visible changes at least twice over the past forty years. As of this writing, its pool is filled with bluish white thermal water with an opalescent character. The blue color is produced by molecules of silica dissolved in the water, which scatter blue wavelengths of light. The length of the dissolved silica molecules ranges between 4 millionths and 200 thousandths of an inch long (0.1 and 0.5 micrometers).

In addition to the dissolved silica are larger, solid silica particles that are not dissolved but suspended in the water. These particles scatter all wavelengths of sunlight and produce the milky appearance

Top: Pearl Geyser in 1969, when its water was relatively clear and bluish, reflecting the presence of small particles of dissolved silica. *Middle:* Pearl Geyser in August 2003, after it had become a fumarole. *Bottom:* Pearl Geyser in 2008, with bluish white water, indicating it contains both small molecules of dissolved silica and larger solid particles of suspended silica. —Top and middle photos courtesy of the National Park Service

of the thermal water. In 1969, Pearl Geyser was roughly the same size and shape as it is today, but its water was clear blue, indicating that the larger particles of solid silica were essentially absent. The color change indicates that somewhere along the way the geyser's plumbing system changed, and today its water contains both the smaller, dissolved silica molecules and larger, suspended silica particles.

Much more striking changes to some of the thermal features at Norris, including Pearl Geyser, occurred in August 2003 during an exceptionally strong geothermal disturbance. The disturbance began unusually early that year when, without warning on March 10, several fumaroles burst out of a forested hill about 2 miles (3.2 km) north-west of the Norris Museum. Within about a month the fumaroles had transformed to hot springs in which thermal water replaced the escap-ing steam. Meanwhile, other changes occurred in Back Basin. By July, water had completely drained away or evaporated from Pearl Geyser, leaving a hissing fumarole. Temperatures 0.5 inch (1.3 cm) below the ground surface in parts of Back Basin rose to 200 degrees Fahrenheit (93 degrees C). Concerned for public safety, the Park Service closed Back Basin on July 23 and didn't reopen it until the fall, after the dis-turbance had ended.

Although the 2003 geothermal disturbance was particularly strong, it did not lead to a hydrothermal explosion. However, a moderately sized hydrothermal explosion did occur during the geothermal dis-turbance of 1989 at Porkchop Geyser (stop 2), which is also in Back Basin, about 150 yards (135 m) from stop 1. Today, Porkchop Geyser is a relatively quiescent boiling spring filled with bluish white, milky thermal water and surrounded by angular boulders ejected during the hydrothermal explosion. The geyser rarely erupts, although it did erupt in July 2003 during that exceptionally strong geothermal disturbance.

As with Pearl Geyser, notable changes have been observed in the behavior of Porkchop Geyser. Before 1971 it was not a geyser at all but an unobtrusive spring with a triangular vent measuring about 1.5 inches (3.8 cm) long and 1 inch (2.5 cm) across. Hot water occasion-ally seeped from this vent, which in those days was called Porkchop Spring. The spring began to erupt infrequently in 1971, with spouts of water about 10 to 15 feet (3 to 4.6 m) high. By 1985 the infrequent eruptions had become a continuous spout of water 20 to 30 feet (6 to

9 m) high, producing a roar that could be heard more than 1 mile (1.6 km) away. This continued for several years until the early afternoon of September 5, 1989, when the height of the spout more than doubled, rising to between 65 and 100 feet (between 20 and 30 m) for a few seconds before the geyser exploded, propelling large and small chunks of rock outward. Some traveled more than 200 feet (60 m). The biggest rock was the size of a refrigerator, measuring more than 6 feet long, 4 feet across, and 2.5 feet deep (1.8 by 1.2 by 0.8 m). The crater that was left after the explosion measures about 38 feet across, 45 feet long, and 4 feet deep (11.6 by 13.7 by 1.2 m).

Top: Porkchop Geyser in 1986, when it spouted a continuous stream of hot water and steam. *Bottom:* The remains of the geyser on September 7, 1989, two days after the hydrothermal explosion. Note the large, striped block of rock to the right of the spring. The stripes are layers of siliceous sinter that precipitated in the geyser's conduit prior to the explosion. —Photos courtesy of the National Park Service

Looking for clues to explain the explosion, scientists from the U.S. Geological Survey and National Park Service examined the geyser within ninety minutes of the event. They found that essentially all of the rocks blasted out of the crater consisted of very dense siliceous sinter that had precipitated as layers on the inside of the geyser's conduit (see vignette 18 for more on siliceous sinter). The layering is still visible in boulders strewn about Porkchop Geyser. On one of the boulders there was a blob of translucent silica that was pliable. When precipitating from thermal water, silica can first form such masses that later harden, so the pliability of the blob indicated that silica had been actively forming on the inner walls of the geyser's conduit prior to the explosion.

Results from long-term monitoring of the geyser's thermal water, which began in 1951, showed that the water temperature and the amount of silica dissolved in the water generally increased until the explosion. The sequence of events that led to the hydrothermal explosion very likely began around 1971, when Porkchop Spring evolved into an occasionally erupting geyser because of siliceous sinter precipitating along the inner walls of its conduit and the increased temperature of water feeding it. By 1985, further increases in water temperature and constriction due to precipitated silica caused the geyser to become a perpetual spouter. The dramatic and sudden change in geyser height just prior to the explosion happened because part of the constriction gave way, allowing more water and steam to flow through the enlarged opening. The rapid release of water and steam produced an immediate and significant decline in pressure in Porkchop's plumbing system, leading to the hydrothermal explosion.

The nearly annual geothermal disturbance at Norris Geyser Basin and longer-term changes in some of its thermal features indicate that the plumbing system in the subsurface is always changing. Through continued monitoring, geologists hope to better understand the conditions that lead to hydrothermal explosions. Of particular concern is the large, deep reservoir of very hot thermal water beneath the basin. If the geothermal disturbance caused the confining pressure in this reservoir to fall sufficiently, it could trigger a massive hydrothermal explosion on par with the largest explosions that have occurred in the park.

Though the 1989 explosion at Porkchop Geyser was an impressive event, it was puny compared to some of the prehistoric hydrothermal explosions that occurred elsewhere in Yellowstone. Eyewitnesses to the 1989 explosion reported that it was over in seconds and very little rock or water was ejected following the initial blast. In contrast, larger hydrothermal explosions may occur in multiple closely spaced blasts as the confining pressure in successively deeper parts of a superheated body of water is reduced. Such explosions can leave behind massive craters.

Indian Pond (stop 3) is an easily accessible crater that formed during a large prehistoric hydrothermal explosion. The pond measures about 550 yards (500 m) across and has steep, sloping sides. A rim of rock debris ejected during the explosion surrounds the pond and stands about 35 feet (11 m) above the surrounding landscape. The rock debris is explosion breccia, a type of sedimentary deposit consisting of angular fragments of all sizes. The explosion hurled out some particularly large rocks, which can be viewed up close along the inner walls of the crater. The rock fragments are angular due to the violence of the fragmentation that occurred during the explosion, and they do not appear to have moved much since they landed.

The average depth of the pond is about 42 feet (13 m), and the main crater contains three smaller craters. Each of these developed during smaller explosions that occurred as decreasing pressure propagated downward. Based on a radiocarbon date derived from a piece of charcoal found in a soil layer beneath the explosion breccia, geologists know that the explosion occurred almost 2,900 years ago. They don't know what triggered it.

After having a look at the rim surrounding Indian Pond, follow the faint footpath southeast from the parking lot to the trees located on the southeast side of the pond. Once you reach the trees, continue across the open meadow, bushwhacking until you reach the edge of

Looking southwest across Indian Pond. Scientists interpret the pond to be the crater of a massive hydrothermal explosion. The berm surrounding the pond consists of fragments of rock that were ejected from the crater during the explosion.

the cliff overlooking the northern shore of Yellowstone Lake. Scramble very carefully down the small gulley that connects the meadow to the shoreline.

Once you are at the shoreline, examine the exposed cliff east of the gulley. You'll notice that the beach gravel makes an unusual crunching sound when you walk on it, like pieces of broken glass. That's because an abundance of glassy fragments are weathering out of a well-exposed, 5-foot-thick (1.5 m) explosion breccia in the cliff above you. The layer is composed of angular rock fragments that are not sorted by size; rather, the fragments accumulated quickly, with no additional transport or sorting by wind, water, or ice. The explosion breccia is not from Indian Pond but from the older and much more massive hydrothermal explosion that formed Mary Bay, the semicircular bay that stretches from the east end of the beach at stop 3 to Steamboat Point, a little more than 2 miles (3.2 km) to the southeast.

Mary Bay is the largest documented hydrothermal explosion crater in the world. Geologists don't know for sure what triggered the hydro-thermal explosion, but some have suggested it was shaking caused by a large earthquake. It's also possible that a tsunami—a series of large waves—caused it.

The Mary Bay hydrothermal explosion breccia exposed in the beach cliff at stop 3. Geologists interpret the layer of sand below the explosion breccia to have been deposited by a tsunami triggered by an earthquake.

Geologists have identified a sand layer that they think is a tsunami deposit, based on its internal structure, just below the explosion breccia visible at stop 3 and elsewhere along the northern shoreline of Yellowstone Lake. During the earthquake, a troughlike section of lake floor in the northwestern part of the lake, only about 3.7 miles (6 km) from the Mary Bay crater, dropped about 3 feet (0.9 m). The sudden drop would have caused lake water to slosh into the trough, possibly creating a tsunami. As the waves moved east toward Mary Bay, they would have increased in size, just as swells in offshore parts of a beach environment become large waves as they move into the shallower surf zone. When the troughs, or low spots, of the waves passed over Mary Bay, the pressure exerted on the superheated water below the lake floor would have been reduced, possibly triggering the hydrothermal explosion. After the tsunami inundated the shoreline and deposited the sand, the fragments of rock ejected during the explosion fell back to Earth, forming the explosion breccia.

Steamboat Point (stop 4) offers a superb view of the Mary Bay hydrothermal explosion crater. Mary Bay, the large curving embayment extending to the northwest from stop 4, is part of the crater, which has a maximum diameter of about 1.7 miles (2.7 km). The northeastern

margin of the crater can be seen in the distance; look northwest for the prominent bench just past the thermal pond, which is on the far (north) side of the road. The southern margin of the crater is in the middle of Mary Bay. The maximum vertical relief of the crater from the rim exposed onshore to its deepest point within the lake is about 345 feet (105 m).

The crater is composed of dozens of smaller craters, which are obscured by Yellowstone Lake. The composite nature of the Mary Bay crater suggests that it formed as a result of multiple individual explosions. Geologists have recognized at least two separate layers of explosion breccia along cliffs on the northern shore of Yellowstone Lake, indicating that at least two large explosions took place. Based on ages of charcoal and a layer of volcanic ash underlying the explosion breccia, and the ages of ancient shorelines that cut it, the geologists estimate that the crater as a whole formed around 13,000 years ago.

Many of the smaller craters have active vents discharging clay-rich thermal water from the lake floor; look for cloudy water in Mary Bay. The clay is derived from volcanic rock the acidic thermal water modifies in the subsurface. Using a remotely controlled underwater vehicle, scientists with the U.S. Geological Survey recorded the temperatures of the fluids leaving these vents. The highest temperatures were around 250 degrees Fahrenheit (120 degrees C), clearly indicating the presence of superheated water. Several prominent fumaroles to the south just below the overview underscore the high temperatures found beneath this part of Yellowstone.

It's possible that both earthquake shaking and a tsunami played a role in triggering the hydrothermal explosions that formed Mary Bay. Regardless, earthquakes remain potential triggers for massive hydrothermal explosions as long as large thermal areas like Norris Geyser Basin and Mary Bay contain huge reservoirs of superheated water. And large earthquakes are all but certain to be a part of Yellowstone's future. In vignette 20, we turn our attention to the biggest earthquake to have struck the Yellowstone region in recorded history.

20.
The Night the Ground Shook
The 1959 Hebgen Lake Earthquake

On the night of August 17, 1959, the sky west of Yellowstone National Park was clear and the air was still. As the last rays of the setting sun faded, the nearly full moon began to rise over Hebgen Lake, a large reservoir at the southern end of the Madison Range, northwest of the park's West Entrance. Since it was the peak of the vacation season, people were camped at various campgrounds along the lake's shores and downstream in the Madison River canyon. Among them was U.S. Geological Survey geologist Irving J. Witkind, who was conducting field studies in the region. Dr. Witkind was staying in a house trailer parked on a small hill at the mouth of Grayling Creek, about 1 mile (1.6 km) northeast of Hebgen Lake. In a second house trailer on the same hill, Mrs. J. B. Epstein was settling in for the night while her husband, Jack, and his nephew were engaged in a late-evening card game.

At 11:37 p.m. local time, without warning, an earthquake struck the region. In a letter she wrote the next day, Mrs. Epstein said:

> All of a sudden the trailer began to shake violently up and down and back and forth. I thought at first that Jack was fooling around and shaking the trailer, but in a split second I looked around and saw: 1. Water pouring out of the wash basin. 2. All dishes, groceries, and clothes falling out of the cabinets. 3. The gasoline lantern hanging from the ceiling, swinging in a 2-foot circle, and looking as if it would fall any minute. If it did it would have set the whole trailer on fire.
>
> There were fantastic rumblings. The farthest thing from my mind was an earthquake. In this same split second I thought that the 100-pound propane tank outside the trailer was starting to

explode and that's what caused the noise and shaking. In pure horror and fright I dashed out the door and screamed for everyone to follow and run as far away from the trailer as possible. Jack was still in the trailer, trying to stop the lantern. He got beaned on the head with it, gave up, and came charging out. He had realized from the first that it was a quake. My complete horror came after I hit the ground and found that it was no better than . . . a glob of jelly. I was frantic—there was nowhere to get away from the fantastic sensation. Jack screamed not to run near the woods because trees were toppling all over. We could hear loud rumblings due to rockslides and landslides in the mountains.

Meanwhile, close by in the other trailer, Dr. Witkind was having a similar experience:

I thought that the trailer had somehow come off its jacks, jumped the chocks, and was rolling down the hill. I scrambled out the front door determined to stop the trailer, no matter what, although I had no idea as to how I would go about it. When I got outside, the trailer was in place, but the trees were whipping back and forth and the leaves were rustling as if moved by a strong wind—but there was no wind. I knew right then that it was an earthquake. I could hear avalanches in the canyons behind me, and could see huge clouds of dust billow out of the canyon mouths.

The Hebgen Lake earthquake, as it came to be known, was the largest single seismic event ever recorded in the Rocky Mountains. Estimates of its intensity range between magnitude 7.1 and 7.5. It severely damaged roads, buildings, and water pipelines in and around West Yellowstone, Montana, and caused a massive landslide as a 1-mile (1.6 km) section of mountainside in the Madison River canyon gave way.

GETTING THERE

Stop 1, the Cabin Creek Campground, is off US 287 13.6 miles (21.9 km) west of its intersection with US 191 and 5.4 miles (8.7 km) northeast of the Earthquake Lake Visitor Center. Park and walk about 100 yards (90 m) north into the woods to see a fault scarp caused by the earthquake. For an interesting perspective on the events surrounding the Hebgen Lake earthquake, proceed to Hebgen Dam, stop 2, 0.6 mile (1 km) southeast of stop 1 on US 287. From stop 2, proceed west on US 287 for 6 miles (9.7 km) to the Earthquake Lake Visitor Center, stop 3. The parking lot of the visitor center, built directly on top of debris from the slide, provides a sobering view of the slide's breakaway zone.

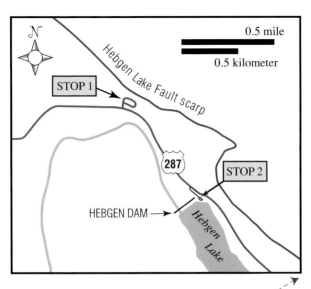

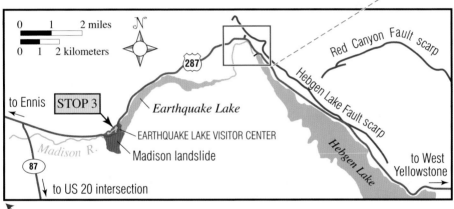

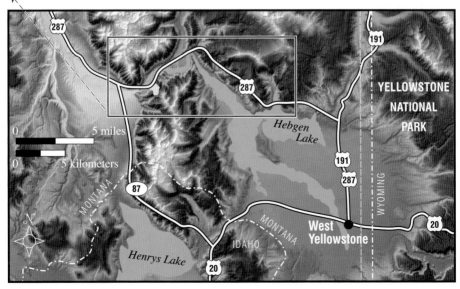

Geophysicists estimate that the earthquake started with the rupture of bedrock about 15 miles (24 km) below the surface and a little north of the town of West Yellowstone. This rupture offset the subsurface bedrock and spread laterally and upward to the ground surface, which itself was ruptured, forming steep embankments called *fault scarps*. The fault scarps stretched more than 16 miles (25.7 km) across the region north of Hebgen Lake and defined the surface locations of the two main faults that broke during the earthquake. The largest scarp had a vertical offset of 21 feet (6.4 m). Many are still plainly visible, although they have been modified by the half century of erosion that has occurred since the earthquake.

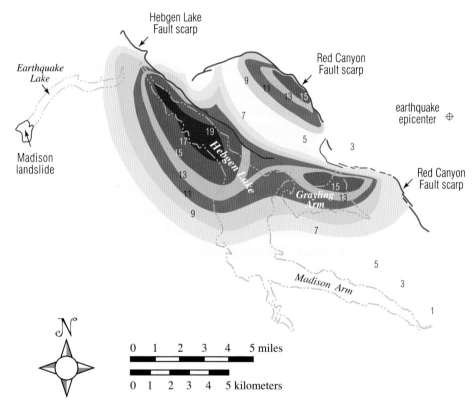

During the 1959 Hebgen Lake earthquake two large fault blocks dropped downward and tilted along major faults. The larger Hebgen Lake Fault block slid downward along the Hebgen Lake Fault and the eastern part of the Red Canyon Fault. The Red Canyon Fault block slid down the western segment of the Red Canyon Fault. The colored contours show how many feet the ground surface dropped.

The two faults that ruptured during the earthquake were the Hebgen Lake and Red Canyon faults. Both are normal faults that project into the subsurface at relatively steep angles. Normal faults develop where the crust is undergoing extension, meaning the crust is being pulled apart. As this tectonic tearing occurs, the block of rock—called a *fault block*—situated above the fault slides downward relative to rock situated below the fault. There are two large fault blocks that slid downward along the Hebgen Lake and Red Canyon faults, and they share the names of the faults.

Stop 1, the Cabin Creek Fault Scarp Area at the mouth of Cabin Creek, offers a convenient view of the Hebgen Lake Fault scarp. From

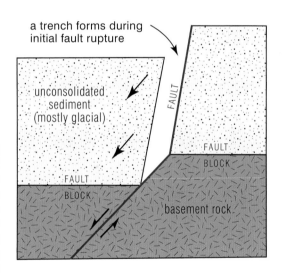

a trench forms during initial fault rupture

unconsolidated sediment (mostly glacial)

FAULT

FAULT BLOCK

basement rock

FAULT BLOCK

Deep, wide trenches formed along some of the scarps during the 1959 earthquake. The trenches formed in the unconsolidated sediment near the surface, where the normal faults were nearly vertical. In the bedrock below the sediment, the faults were not as steep, so no horizontal separation occurred. The biggest trenches were up to 20 feet (6 m) deep and 5 feet (1.5 m) wide. They were not stable features, and the sediment overhanging the trenches slowly collapsed into them during the months and years following the earthquake.

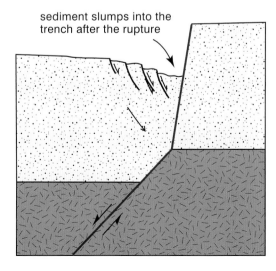

sediment slumps into the trench after the rupture

the parking area, walk up the short trail to the prominent dirt bank—the scarp. This portion of the scarp is the western end of the Hebgen Lake Fault. When the fault ruptured, it broke through the middle of the Cabin Creek Campground, forming a dirt wall 20 feet (6 m) high that cut off the north end of the campground from the main road and temporarily stranded a number of campers. The ground you are standing on is part of the Hebgen Lake Fault block.

It would be devastating if one of these fault scarps formed in a major city. Streets would end against a dirt wall, from which the severed ends of water mains, sewage pipes, electrical cables, and so forth would protrude. Tall buildings would lean over as the ground beneath them warped. Within the United States, Salt Lake City is probably at the greatest risk for such a catastrophic event, since it is built along the Wasatch Fault at the base of the Wasatch Range. This fault is a major normal fault that has ruptured at least four times in the past 10,000 years, most recently about 1,300 years ago.

From Cabin Creek Campground, you are about 0.6 mile (1 km) downstream from Hebgen Dam (stop 2). Built between 1909 and 1911, Hebgen Dam consists of an embankment made of clay, sand, and gravel on the lake side; a concrete-core wall buried beneath the service road

The fault scarps produced by the 1959 Hebgen Lake earthquake were impressive because they were tall and steep. In places fissures formed, such as the one in which the geologist is standing. —Courtesy of J. R. Stacy, U.S. Geological Survey

running the length of the dam; and an embankment made of loose rock on the downstream side. The dam is situated on the Hebgen Lake Fault block and dropped about 9 feet (2.7 m) during the earthquake. It was significantly damaged due to ground shaking. At the northeastern end of the dam, closest to the spillway, the lake-side embankment settled as much as 7 feet (2.1 m) and the downstream side settled nearly 4 feet (1.2 m). Both embankments also expanded laterally. As a result, the concrete core protruded above the embankments once the shaking had finally stopped. The core wall cracked in at least sixteen places and was bent as much as 3 feet (0.9 m) from its original straight orientation.

Hebgen Lake has a surface area of about 21 square miles (54 square km)—about a fifteenth the surface area of Yellowstone Lake. When the earthquake struck, the floor of the lake basin dropped as the Hebgen Lake and Red Canyon faults ruptured. Detailed leveling surveys and depth soundings conducted after the earthquake showed that the lake floor and shoreline dropped less than 1 foot (30 cm) at the head of the lake but more than 21 feet (6.4 m) at its foot, near the dam. This sudden dropping and tilting caused water to flow toward the foot of the lake, where the drop was greatest. Water overran the shoreline, stopped,

The Hebgen Lake Fault scarp at Cabin Creek Campground in 2008, nearly half a century after the earthquake. The tree on the downdropped side of the scarp was tilted by the movement along the fault but continued to grow vertically, giving its trunk a curved shape.

The Red Canyon Fault scarp passed near several farm outbuildings and directly under this building, which collapsed completely. —Courtesy of I. J. Witkind, U.S. Geological Survey

and then flowed back across the lake to the opposite shoreline before returning again to the foot of the lake. This type of slow oscillation of lake water is called a *seiche.*

George Hungerford, a Montana Power Company foreman who was living in company quarters at the Hebgen Dam, provided a very interesting personal account of the seiche. Once the quake had ended, Hungerford and an assistant went to check on the dam.

> The dust was so intense you could hardly see. You could hardly breathe, or anything. It obscured the moon. I walked over to the edge of the dam and all we could see was blackness. There was no water. No water above the dam at all, and I couldn't imagine what had become of it. By that time the dust had started to clear, and the moon had come out a little. And then here came the water. It had all been up at the other end of the lake. We rushed back when we heard the water coming. We could hear it before we could see it. When it came over the dam, it was a wall of water about three to four feet high completely across that dam, and it flowed like that for what seemed to me to be 20 minutes, but possibly it could have been 5 or 10. I have no idea of time. It flowed for a while, and then it started to subside. Then it all cleared away, and no water again. The lake was completely dry as

far as we could see. All we could see down the dam was darkness again. It seemed like a period of maybe 10 to 15 minutes, and the water came back, and then it repeated the same thing over again. It did that the third time, but the third time was nothing like the first two times. . . . Then it kept surging back and forth, but it never went over the dam again.

The water rolled back and forth for more than eleven hours after the earthquake, although with each trip across the lake the size of the seiche diminished. In addition to overtopping Hebgen Dam at least three times, the water swept past the north shoreline of the lake, stranding debris well onshore. Three separate sloshes of the seiche left separate debris lines above the shoreline near Hebgen Dam. The highest debris was stranded about 8 feet (2.4 m) above the lake surface. In addition to causing the seiche, the dropping land triggered several small landslides and caused sections of US 287 to fall into Hebgen Lake, along with an entire house. Overall, about 50 square miles (130 square km) of land dropped more than 10 feet (3 m) during the quake, and roughly 200 square miles (520 square km) subsided more than 1 foot (30 cm).

The possibility that Hebgen Dam might fail and release a giant flood was a significant concern for residents living downstream along the Madison River, especially in the town of Ennis, Montana, about 50

This house, originally located on the northern shore of Hebgen Lake, slid into the reservoir, along with parts of US 287, as a result of the earthquake. —Courtesy of J. B. Hadley, U.S. Geological Survey

miles (80 km) below the dam. Within a few hours of the earthquake, around 2:45 a.m. on August 18, a message was received that Hebgen Dam had failed or was about to fail. Authorities sounded sirens, and people went house to house to warn friends and neighbors. By 4:00 a.m. Ennis was nearly deserted, most residents having gathered in a large, high field outside of town. By dawn the field was covered with cars. After the sun came up, many residents started returning home. Later in the morning, however, the Montana Power Company reported that the top of the dam had been significantly damaged, so town authorities issued an official evacuation order. The order was retracted on the afternoon of August 19, and residents were allowed to return to their homes on a standing alert.

Although in the confusing aftermath of the quake many thought the dam had failed or was about to, it never did. This is remarkable because the Hebgen Lake Fault scarp passes within 0.25 mile (0.4 km) of the dam, and it was movement along this fault that seriously warped the floor of the lake, caused the seiche, and damaged the dam. By far the most significant actual disaster associated with the earthquake occurred about 6 miles (9.7 km) downstream of the dam, near stop 3.

When the earthquake struck, campers at the Rock Creek Campground in the Madison River canyon below Hebgen Lake had mostly fallen asleep for the night, including Mr. and Mrs. F. R. Bennett and their four children. Shortly after falling asleep, Mrs. Bennett awoke to a loud noise, followed "some time later" by a great roar. Alarmed, she and her husband left their small house trailer to check on the children, who were sleeping in bedrolls on the ground nearby. As they left the trailer, they were struck by a huge blast of air. Mrs. Bennett saw her husband grab onto a tree. The air blast lifted him off his feet and strung him out "like a flag" before he let go. Before losing consciousness, she also saw one of her children blow past her, followed by a car tumbling end over end. Her sixteen-year-old son, Phillip, was blown about by the wind, broke his left leg, and was immersed in water from the Madison River, which was pushed out of its channel by a landslide. He survived the night by crawling into a clump of trees and covering himself with mud to stay warm. The next morning he and Mrs. Bennett, having lived through what became known as the Madison landslide, were rescued from the campground. The rest of their family had perished.

Eyewitnesses of the Madison landslide all agree that a few seconds after the earthquake there came a tremendous roar. Air Force Warrant Officer Victor James was in a trailer parked in the Rock Creek Campground about 75 yards (70 m) from where the main slide struck. Afterward he said, "I heard a terrible rumble and looked up. I saw the whole damn mountain crumbling. It was awful. I saw a lot of fighting during World War II, but I never heard such a hell of a roar."

The Madison landslide is one of the three largest landslides to have occurred during historic time in North America. A 1-mile-long (1.6 km) section of the southern wall of the Madison River canyon failed, releasing an estimated 28 million cubic yards (21 million cubic meters) of rock. The rock making up the slide broke apart on its way down, increasing its overall volume by about 32 percent (due to air space between the fragments once they came to rest), so the total volume of the slide is estimated to be about 37 million cubic yards (28 million cubic meters). The Madison landslide is about the same size as the 1903 Turtle Mountain landslide at Frank, Alberta, estimated to be between 35 and 40 million cubic yards (27 and 31 million cubic meters), and somewhat smaller than the 1925 Gros Ventre landslide in northwestern Wyoming, estimated at about 50 million cubic yards (38 million cubic meters).

From the parking lot of the Earthquake Lake Visitor Center (stop 3), the upper part of the slide scar is clearly visible. It extends from the base of the steep slope of exposed rock (directly across the canyon) all the way to the skyline. Although debris from the slide covers the base of the scar, geologists estimate that a section of the canyon wall roughly 1,000 feet (300 m) tall broke away and slid downhill. The earthquake literally shook off the top of the mountain!

The landslide occurred in Precambrian rocks that were uplifted and tilted around 65 million years ago as the mountains in the area were rising (see vignettes 5 and 6). At the bottom of the canyon wall was dolomite, a limy sedimentary rock deposited in an ancient sea. Layers of the dolomite dipped toward the river at a steep angle. Above the dolomite were schist and gneiss with foliation—layering formed during metamorphism—that also dipped steeply toward the river. The large, flat rock surfaces sloping toward the canyon and located at the top of the slide scar are foliation planes. Because dolomite is more resistant to weathering than schist and gneiss, it had formed a small but discontinuous ridge that acted like a buttress holding back the weaker layers of schist and gneiss above. During the earthquake, ground shaking caused the dolomite buttress to fail, and the layers of schist and gneiss slid downward along their steeply tilted foliation planes. The light-colored rock outcropping in the lower right (western) portion of the slide scar is a remnant of the dolomite buttress. The rest of the slide scar consists of darker schist and gneiss.

The rocks and rock piles at the level of the highway in front of you, around you, and behind you slid downward as much as 1,385 feet (422 m)—the distance from the former mountaintop to the elevation of the Madison River prior to the slide. The landslide blasted across the Madison River and more than 400 vertical feet (122 m) up the

The Madison landslide as viewed from stop 3. The landslide traveled downhill directly toward you. The boulders and other rock debris in the foreground are part of the slide. The outcrops of light-colored rock are remnants of the dolomite buttress that collapsed to initiate the slide.

canyon behind you. Geologic mapping of the slide has shown that it moved downslope as one giant, semi-intact mass, so fragments of the broken dolomite buttress formed the leading edge of the slide. Today, these fragments occur mostly on the north side of the slide mass. Most of the rest of the slide mass consists of bits and pieces of schist and gneiss that followed the dolomite downslope.

As recounted by survivors from Rock Creek Campground, the slide created a terrific wind that tore the clothing off campers and blew parked cars across the campground. Although some of the wind was simply air pushed along in front of the massive slide, the strongest wind was probably related to the mass having partially launched into the air. As the slide mass ran past the remnants of the buttress, it became airborne, and air trapped under it was forced outward laterally as the mass landed in the canyon bottom.

Air wasn't the only thing forced outward from the slide. As the mass of rock surged across the Madison River, it forced water out of the river channel. Violently displaced water formed a wave that carried trees, driftwood, and small rocks both upstream and downstream, and over both banks of the river channel. The wave reached between 0.25 and 0.5 mile (0.4 to 0.8 km) upstream of the slide and crested at a maximum height of about 100 feet (30 m) above the river. It's this wave that engulfed Rock Creek Campground, which was located on the north bank of the river. Downstream of the slide, the wave crested at least 15 feet (4.6 m) above the river, stranding logs and fish in the sagebrush along the riverbank and sweeping two cars 100 yards (90 m) past the slide's lower edge.

The Madison landslide involved dark schist and gneiss that display obvious layering related to foliation, and more-massive dolomite. The hammer is sitting on schist, and the large, lighter-colored boulder is dolomite.

When it finally came to rest, the slide completely blocked the Madison River, forming a natural dam. Earthquake Lake began to form as the river backed up and water overtopping Hebgen Dam arrived. The slide buried the western third of Rock Creek Campground, and Earthquake Lake inundated the rest of the campground almost immediately after the earthquake. By 6 a.m., all the cars in the campground were submerged beneath the water of the rising lake. Twenty-six people went missing and are presumed buried under the slide. The U.S. Forest Service never reopened the campground.

Fearing that Earthquake Lake would grow large enough to cause the slide dam to fail, releasing a giant flood, the U.S. Army Corps of Engineers excavated a spillway 200 feet (61 m) wide by 0.5 mile (0.8 km) long across the top of the slide to keep the level of the lake as low as possible. On September 10, twenty-four days after the slide, water from Earthquake Lake began flowing through the spillway. Eventually, about half of the water in the lake was drained. Today, dead trees that once lined the banks of the Madison River stick out of the ghostly body of water. Indeed, it is quite a sobering experience to look upon the scene and ponder the events that took place that moonlit and windless summer night in 1959, when an earthquake struck, a mountainside moved, and the shape of Yellowstone Country changed forever.

Aerial view of the Madison landslide, looking downstream to the southwest, a few days after the slide. The slide dammed the Madison River, and Earthquake Lake, in the foreground, began to form immediately. —Courtesy of J. R. Stacy, U.S. Geological Survey

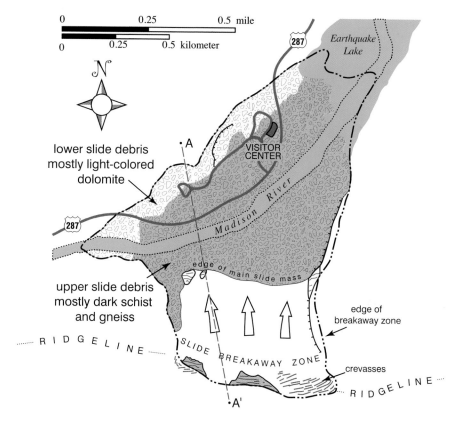

Map and cross section (A-A') of the Madison landslide as it looked in 1959. The slide was nearly 0.5 mile (0.8 km) wide and more than 1 mile (1.6 km) long and moved from the southeast to the northwest. Very shortly after the Hebgen Lake earthquake, the low, ridge-forming outcrops of Precambrian dolomite overlooking the Madison River collapsed, removing the support for the mountainside of weak gneiss and schist above and initiating the slide. The leading edge of the slide was dominated by dolomite debris from this bedrock ridge.

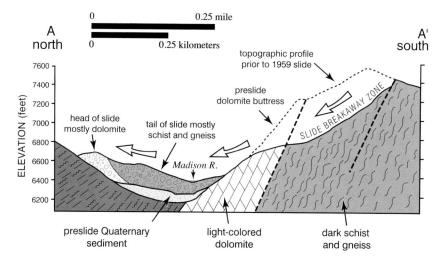

Epilogue

The Certainty of Change in Yellowstone

In 1988 the largest complex of wildfires since the establishment of Yellowstone National Park in 1872 swept through the region. Starting in early July, the fires burned until early October, when rain and early snow stopped their spread. One-third of Yellowstone National Park burned, more than 2,000 square miles (5,180 square km)—an area larger than the state of Delaware. On August 20 alone, as a dry wind raked across Yellowstone Country, the fires burned across 150,000 acres, about 234 square miles (606 square km). The Park Service was forced to close the entire park to nonemergency personnel. Today, more than twenty years after the fires, abundant snags of burned trees from the 1988 wildfires are still visible across much of the region.

The cost of fighting these wildfires was staggering—more than $120 million. Never before in the history of the United States had more been spent fighting fires. A total of about twenty-five thousand firefighters participated directly in the effort to control the flames in and around the park, including six hundred U.S. military personnel. More than 600 miles (965 km) of hand-dug fire line were cut in an attempt to slow the flames, an effort that, in general, failed. More than 1 million gallons (3.8 million liters) of fire retardant were dropped on the fires, in addition to more than 10 million gallons (38 million liters) of water—enough to fill fifteen Olympic-sized swimming pools.

The immense size of the wildfires resulted from a number of factors. The six years prior to 1988 had been unusually wet, promoting the growth of abundant vegetative fuel. During these years, few fires of any size occurred within the park. The spring of 1988 also was quite wet: Precipitation levels in April and May were 143 percent and 188 percent of average, respectively. By the end of May, however, the spring precipitation had all but ended, and by mid-June dry winds and a lack of rain combined to create droughtlike conditions. Humidity less than 20 percent was common, promoting the evaporation of moisture from surface water, vegetation, and the soil.

One of the reasons the forests dried so rapidly is the granular character of soil across Yellowstone. Much of the soil is derived from weathered rhyolite bedrock composing the lava flows that filled in the Yellowstone Caldera (discussed in vignette 13). When rhyolite weathers, it breaks down into pea-sized granular chunks that have little nutrient content, clay, or organic material that would help the soil retain moisture. The porous nature of the soil also allows water to be wicked relatively easily toward the surface, where it evaporates during periods of sustained low humidity.

During the sixteen years prior to the 1988 fires, a fire policy had been in effect in which natural fires were allowed to burn in Yellowstone National Park, provided they posed little threat to property, while fires outside the park were fought. Many ecologists and government land

Much of the soil in Yellowstone National Park is derived from the weathering of rhyolite bedrock, which produces a granular texture not capable of holding much water. Nickel for scale.

managers viewed the hands-off firefighting policy, which recognizes fire as an important part of the ecosystem, as a success because no major fires had occurred in the park since the policy had been instituted. Since 1988, though, large, uncontrollable fires have become much more common in the American West, leading to a vigorous debate over the nation's firefighting policy, including in Yellowstone. In large measure, this debate is tied to the issue of global warming. As more and more evidence in support of global warming has been discovered, many have begun to wonder how the warming will affect the frequency and behavior of large fires.

Scientists studying sediment cores collected from small lakes and ponds in the Yellowstone region have provided valuable insights into this issue. When a wildfire occurs, some of the resulting charcoal gets washed into lakes and ponds and accumulates along their bottoms with other sediment, such as small mineral grains or the remains of plants. The sediment provides a record of how a landscape has varied through centuries and millennia, including information about the sorts of plants

Looking northeast toward Bunsen Peak from Grand Loop Road. The snags are trees that burned in September 1988. The smaller, live trees have established themselves since the fires. Evidence of the 1988 wildfires is clearly visible around much of Yellowstone National Park.

and animals that thrived in the region. Scientists can also estimate the frequency of fires in the region by studying these layers.

Some of the best records of fire frequency in Yellowstone, spanning the last 17,000 years, approximately the period of time since the Pinedale glaciation, come from the Cygnet Lakes. These small lakes are located on rhyolite flows within the Yellowstone Caldera, south of the road between Norris Geyser Basin and Canyon Village. Geoscientists studying sediment cores from the lakes estimate that between about 17,000 and 9,900 years ago, the number of fires increased from fewer than 5 to about 15 fires per 1,000 years. After 9,900 years ago, the number of fires per 1,000 years steadily decreased, so that by 2,000 years ago only 5 fires occurred every 1,000 years, and only 2 or 3 fires per 1,000 years have occurred since.

The greater number of fires around 9,900 years ago was the result of a warmer climate. At that time the Earth's axis was tilted slightly more than it is now, causing the sun to shine more directly overhead during the northern hemisphere summer and to sink closer to the horizon during the northern hemisphere winter. In addition, the point in Earth's closest orbit to the sun—called *perihelion*—occurred during the northern hemisphere summer. Today, this closest approach takes place during early January. The combination of a greater tilt angle and a summertime perihelion caused about 8.5 percent more sunlight to reach the northern hemisphere during its summer relative to today. The increased solar energy brought about greater levels of warmth and aridity that led directly to more frequent fires. In contrast, the global warming that is occurring today is the result of increased amounts of greenhouse gases in the atmosphere. These gases absorb infrared radiation from the sun and reemit it, causing the atmosphere to warm up.

So what effects will global warming have on fire frequency in Yellowstone? Scientific models suggest that a doubling of our current carbon dioxide concentrations over the next several centuries will warm the atmosphere in the Yellowstone region between about 3 and 7 degrees Fahrenheit (about 1.5 and 4 degrees C). The last time a temperature rise of this magnitude occurred was between 14,000 and 9,500 years ago, when glaciers receded across the Yellowstone region, transforming the landscape from tundra to forest. Based on comparisons with data from this period of increased fire frequency, scientists studying the Cygnet Lakes cores have suggested that fire frequency will likely increase in Yellowstone as a result of global warming. Fortunately, it takes time for forests to grow back after a fire, so future fires may not have the large amount of fuel that was available across the region in 1988, after several decades of uninterrupted growth and several especially wet years. Indeed, scientists concluded that although more fires were occurring roughly 9,900 years ago, they were probably smaller because they reduced and limited the overall amount of fuel.

While the potential drying of the Yellowstone region may affect the frequency and behavior of wildfires, it may also change the behavior of the park's thermal features. A drying climate could lead to long-term drought, which could cause the water table to drop. Perhaps the largest hazard that might result from long-term lowering of the water table is a hydrothermal explosion. Like the sudden, violent explosion of Porkchop Geyser (discussed in vignette 19), these occur when superheated groundwater flashes to steam due to some sort of pressure release, such as the lowering of the water table. These explosions can hurl large rocks long distances. Several large boulders, including one the size of a small refrigerator, were thrown from Porkchop Geyser, and debris was hurled 200 feet (60 m). Much larger hydrothermal explosions have occurred in Yellowstone in the past, leaving craters that include Indian Pond, about 550 yards (500 m) across, and Yellowstone Lake's Mary Bay, about 1.7 miles (2.7 km) across.

How likely is it that a global warming will cause hydrothermal explosions in Yellowstone National Park? Although a lot more information is needed to answer this question accurately, the annual geothermal disturbance at Norris Geyser Basin indicates that this area, in particular, appears to be sensitive to relatively small seasonal fluctuations in the water table. As discussed in vignette 19, the annual disturbance at Norris turned violent in 2003 when new fumaroles burst out of a nearby hillside and unusually hot ground temperatures were recorded across parts of the basin. Although no one knows whether global warming will cause the water table under Norris Geyser Basin to drop sufficiently to cause a major hydrothermal explosion, scientists with the Yellowstone Volcano Observatory are carefully monitoring the temperature and amount of water discharging from hot springs and geysers in the basin because this data can serve as an indicator of changing thermal conditions that might precede an explosion.

Regardless of global warming, Yellowstone will certainly experience violent change related to the massive Yellowstone hot spot. As the North American Plate continues tracking southwestward over the hot spot, ground passing over the hot spot's leading edge will rise while the ground in its wake will subside. These vertical motions of the crust will be accommodated by major faults that already exist, and new faults will form as the crust is fractured. Across the northern Rockies, there are hundreds of active faults interpreted to be influenced by the tectonic activity associated with the hot spot. Some will undoubtedly produce large earthquakes similar to the 1959 Hebgen Lake earthquake (discussed in vignette 20).

Although wildfires, hydrothermal explosions, and earthquake-generating fault ruptures are all likely to be part of Yellowstone's future, these hazards pale in comparison to the catastrophic effects that a major caldera-forming eruption of the Yellowstone Volcano would have on Earth's inhabitants. Fortunately, despite the volcano's highly active nature, a caldera-forming eruption is not imminent. However, about 760,000 years passed between Yellowstone's first and second caldera-forming eruptions, and about 650,000 years separated the second eruption from the third, which was the most recent. About 640,000 years have elapsed since the most recent eruption. These time intervals suggest that the next major eruption could take place relatively soon—in terms of geologic time.

Despite the various signs indicating that the Yellowstone Volcano may be due for another eruption, it probably won't take place within our lifetimes. Instead, the millions of annual visitors to Yellowstone Country are more likely to see smaller-scale changes in the region's thermal features, landscapes, and ecosystems. For current and future generations, changes in this unique international treasure are all but certain, presenting a one-of-a-kind opportunity to witness firsthand an amazing geological system as it transforms, literally, under our feet.

Glossary

abrasion. The mechanical wearing or grinding away of rock surfaces by wind, flowing water, or ice.

Absaroka Volcanic Supergroup. An extensive, thick sequence of volcanic rocks, sedimentary rocks, small plutons, and shallow intrusive igneous rocks related to multiple large volcanoes that existed in Yellowstone Country during Eocene time.

advance (glacier). Forward movement of the downstream end of a glacier caused by net seasonal (annual) accumulation of snow, which ultimately transforms to glacial ice, or by changes in glacial dynamics that lead to forward movement.

amphibole. A relatively common, dark silicate mineral characterized by prismatic crystals.

andesite. A fine-grained, dark, primarily extrusive igneous rock that is richer in silica than basalt but not as silica-rich as rhyolite.

anion. A negatively charged ion.

ash. See **volcanic ash**

ash-fall tuff. A well-layered, loosely packed jumble of small pieces of volcanic rock and ash that fall back to earth following the initial phase of an explosive volcanic eruption. Ash-fall tuffs commonly are overlain by welded tuffs produced by the collapse of an eruptive column that develops after the initial eruption.

ash flow. Synonymous with **pyroclastic flow**. A hot mixture of volcanic ash, small bits of magma, and gas that flows rapidly down the flanks of a volcano or along the Earth's surface as a result of an explosive volcanic eruption.

ash-flow tuff. Volcanic ash and pieces of volcanic rock deposited by a pyroclastic flow. Includes both nonwelded and welded tuff. Commonly, a single ash-flow tuff is welded at lower stratigraphic levels but remains unwelded at higher stratigraphic levels.

asthenosphere. The portion of the upper mantle beneath the lithosphere, about 60 to 215 miles (100 to 350 km) below Earth's surface; consists of weak, plastic rock.

basalt. A fine-grained, dark, primarily extrusive igneous rock that is relatively rich in calcium, iron, and magnesium and relatively poor in silicon.

basement rock. Igneous and metamorphic rock that underlies younger, layered sedimentary rock.

Beartooth Plateau. The high, relatively flat topography found in the heart of the Beartooth Mountains; characterized by widespread exposure of uplifted igneous and metamorphic rock of Precambrian age that has been deeply eroded by glaciers.

bedding. The layered structure of sedimentary rocks formed during sediment deposition and caused by vertical variations in sediment size, composition, or sorting.

bedrock. The solid, unbroken rock that underlies loose material, such as soil and gravel.

beds. The layers in sedimentary rocks formed by vertical variations in color, sediment size, and other sedimentary characteristics. Beds range from as thin as 0.25 inch (6 mm) to more than 10 feet (3 m) thick.

biotite. A common rock-forming silicate mineral, usually black or dark brown, and characterized by cleavage that allows it to flake into thin sheets; sometimes occurs as hexagonal crystals.

boulder. A rock fragment that is larger than 25.6 cm (about 10 inches) in diameter.

breccia. A sedimentary rock consisting of angular rock fragments held together by a fine-grained matrix or mineral cement.

butte. A flat-topped mountain with several steep cliff faces.

calcite. A widespread, abundant mineral composed of calcium carbonate. Calcite is the major component of limestone, including travertine.

calcium carbonate. A white, crystalline compound found in chalk, limestone, and many shells.

caldera. A large, circular or elliptical basin that forms where the ground collapses following a very large volcanic eruption.

cirque. A steep-walled, semicircular depression eroded into a mountainside by a glacier.

cirque glacier. A glacier that erodes and occupies a cirque.

clast. An individual grain or fragment of a sedimentary rock that was eroded from a larger preexisting rock mass.

clay. An individual mineral particle of any composition having a diameter less than 4 microns. Also, an earthy, extremely fine-grained, highly plastic deposit composed primarily of clay-sized particles.

cleavage. Planes of weakness along which minerals break.

climate. The characteristics of weather, including temperature and precipitation, averaged over a significant period of time, on the order of decades to centuries.

cobble. A rock fragment with a diameter between 6.4 and 25.6 cm (roughly 2.5 and 10 inches).

columnar joints. Parallel, prismatic columns, polygonal in cross section, which can develop in intrusive igneous rocks that cool near the surface and in extrusive igneous rocks; particularly common in basalt but also form in welded tuffs. The joints develop as the rock mass cools and contracts.

compression. Tectonic stress that shortens an object by squeezing it.

conglomerate. A sedimentary rock consisting of rounded pebbles, cobbles, or boulders held together by a fine-grained matrix or mineral cement.

contact. The surface separating two different rock bodies.

cosmogenic surface exposure dating. A geologic dating technique that determines the amount of time a particular rock has been resting at Earth's surface.

crack. A break in a rock along which there is not complete separation of the two sides.

crater. A steep-walled, circular or semicircular depression generally caused by a volcanic eruption or hydrothermal explosion.

crossbed. An arrangement of granular sediment in which strata are inclined at an angle to the main stratification; caused by sediment deposition by a moving fluid, such as river water or wind.

crust. The Earth's outermost layer of rock, composed of relatively low-density, silicon-rich rocks.

crystal. A solid element or compound in which the atoms are arranged in an orderly and periodically repeating array that may be expressed by planar faces.

debris flow. A moving mass of rock fragments, soil, and mud. Debris flows can travel as slowly as 3 feet (1 m) per year and as fast as 100 miles per hour (160 kph).

deep time. The concept of geologic time.

deformation. Folding, faulting, and other changes in the shape of rocks, minerals, or other substances, including ice, in response to mechanical forces, such as those that occur in tectonically active regions, or in flowing glacial ice.

delta. A sediment accumulation that develops where a stream empties into a standing body of water, such as a lake or ocean.

dike. A sheetlike igneous rock that cuts across the structure of preexisting surrounding rock.

dip. The angle formed between horizontal and the plane of interest, measured parallel to the direction a marble would roll if placed on the plane.

discharge. The volume of fluid moving past a particular point over a given unit of time; for a stream or river it is usually measured in terms of cubic feet per second or cubic meters per second.

divide. A topographic high point that separates surface runoff into two different drainage basins.

earthflow. A mass-movement process and landform characterized by the downslope sliding and flowing of soil, weathered rock, and mud. Earthflows typically have well-defined lateral boundaries and terminate in lobelike forms.

earthquake. A sudden motion or trembling in the Earth and at the Earth's surface caused by the abrupt release of accumulated strain.

erosion. The removal of weathered rock, minerals, and soil by moving water, wind, ice, and gravity.

erratic. A boulder that was transported to its present site by a glacier and deposited some distance from its source. An erratic generally is made of a different rock type than that on which it rests.

eruptive column. A roughly vertical column of ash, hot gas, and minor amounts of magma extruded from a volcano during an explosive eruption.

extension. Tectonic stress by which rocks are pulled apart.

extrusive rock. An igneous rock formed from magma erupted onto the surface of the Earth.

fault. A fracture in rock along which rock on one side has moved relative to rock on the other side. See also **reverse fault, thrust fault**

fault block. A unit of crust partially or completely bounded by faults that moves as a single unit during tectonic deformation.

fault scarp. An escarpment formed by the rupture and offset of the ground surface during movement along a fault.

feldspar. A group of common rock-forming minerals composed mostly of silica and aluminum, plus calcium, sodium, and/or potassium.

fissure. An extensive crack, break, or fracture in rock.

floodplain. A low-lying region adjacent to a river channel and subject to periodic flooding.

foliation. Layering in a metamorphic rock formed during the metamorphic process. Foliation is characterized by changes in the size, shape, and orientation of mineral crystals.

formation. A formally recognized body of rock that consists dominantly of a certain rock type and that is distinct and mappable.

fracture. A break in rock or minerals, excluding mineral cleavage. Includes irregular, nonplanar breaks; planar, nonparallel cracks; and regularly spaced, planar, parallel cracks or joints.

fumarole. A vent or hole in the ground from which steam, gases, and/or vapors issue.

geyser. A hot spring that intermittently erupts jets of hot water and steam. A geyser forms when groundwater comes into contact with rock that is hot enough to create steam under conditions preventing rapid escape.

geyser basin. A valley that contains numerous hot springs, geysers, and fumaroles fed from the same reservoir of groundwater.

glacial period. A part of geologic time in which glaciers covered a much larger total area than those of the present day.

glacial till. See **till**

glaciation. The formation, advance, and recession of glaciers. The geologic processes of glacial activity and their resulting effects on Earth's surface.

glacier. A large mass of ice that forms on land by the accumulation, compaction, and recrystallization of snow and that either moves downslope or spreads outward through a combination of internal flow and slipping along its base.

gneiss. A strongly metamorphosed rock characterized by layers, or foliation, resulting from changes in the size, shape, and orientation of minerals.

granite. A plutonic igneous rock made predominantly of feldspar and quartz and commonly containing mica; a major rock type in continental crust.

ground moraine. Till that accumulates at the base of a glacier. Landscapes composed of ground moraine typically are hummocky.

group. A formally recognized body of rock ranking above formation and below super-group. A group contains two or more formations.

hot spot. A volcanic center, typically 60 to 125 miles (100 to 200 km) in diameter and persisting for millions of years; commonly thought to be the surface expression of a rising plume of hot mantle rock.

hot spring. A thermal spring with a temperature that is higher than that of the human body (98 degrees Fahrenheit, or 37 degrees C).

hydrothermal explosion. The violent ejection of water, steam, rock, and mud from the ground caused by the release of pressure and the flashing of superheated water to steam; hydrothermal explosions generally form craters.

ice age. A loose synonym for **glacial period**, especially those that occurred during Pleistocene time.

ice cap. A dome-shaped or platelike covering of glacial ice and snow that covers the summit regions of a mountain range or continent and flows outward in all directions.

ice divide. A subtle topographic ridge on a large ice mass, such as an ice cap, separating ice flowing in different directions and ultimately to different outlet glaciers.

ice stream. A current of fast-moving glacial ice within a larger ice cap; usually ice streams flow down mountain valleys.

igneous rock. Rock that solidifies from magma.

intrusive rock. An igneous rock formed from magma that is injected into and cools within bodies of preexisting rock.

ion. An atom that has gained or lost electrons and has a negative or positive charge. See also **anion**

joint. A rock fracture across which no relative movement occurs; typically occurs as a series of regularly spaced, parallel, planar cracks.

landslide. A general term for rapid downslope movement of soil and rock, including rockfalls, slumps, earthflows, and mudflows. Landslides can result in a variety of landforms.

lateral moraine. An accumulation of glacial till that forms along the lateral margins of a glacier and is left behind as a ridge or embankment after the glacier retreats.

lava. Fluid magma that flows over Earth's surface from a volcanic eruption and cools to form volcanic rock.

lava flow. An outpouring of lava from an erupting volcano or vent; also used to describe the volcanic rocks formed by the cooling and solidification of such an outpouring.

layer. A synonym for **bed**.

layering. A synonym for **bedding**.

limestone. A sedimentary rock consisting mainly of calcium carbonate.

lithosphere. That part of Earth, including the crust and uppermost mantle, that rests above the asthenosphere.

lithospheric plate. A relatively rigid segment of the lithosphere that moves independently of surrounding parts of the lithosphere; also called *plate*. See also **plate tectonics**

magma. Molten rock generated within the Earth.

magma chamber. A reservoir of molten rock within the lithosphere; usually characterized by an interconnected, three-dimensional network of liquid magma within otherwise solid rock rather than an underground reservoir that is purely fluid.

mantle. A mostly solid layer of Earth lying below the crust and above the outer core. The mantle is about 1,780 miles (2,870 km) thick and made of very dense rock that flows very slowly over geologic time.

massive. Said of rocks that have a relatively homogeneous texture and lack layering, fracturing, or other discontinuities.

matrix. Finer-grained material in between and/or surrounding larger grains of sediment in sedimentary rock; the small crystalline minerals in an igneous rock, including noncrystalline glass in volcanic rock; the sediment or rock in which a fossil is embedded.

meltwater. Water formed from the melting of snow and ice, including glacial ice.

member. A body of rock that ranks below formation and may or may not be formally defined or mappable. Members can extend from one formation to another.

metamorphic rock. Rock that has undergone metamorphism.

metamorphism. Changes in the mineralogy and texture of rocks resulting from exposure to high temperature and pressure deep within the Earth. Most changes occur while the rock is solid.

mineral. A naturally occurring, crystalline, inorganic solid with a characteristic chemical composition and atomic structure.

mold. A fossil in which the impression of a plant or animal is preserved in rock.

moraine. A distinct ridge or series of ridges composed of glacial till and deposited by glacial ice. See also **ground moraine**, **lateral moraine**, **recessional moraine**, **terminal moraine**

mud. Wet silt and clay.

mud pot. A hot spring containing boiling mud, typically sulfurous, and may be multicolored. The mud in most of Yellowstone's mud pots formed as acidic thermal water broke down the Lava Creek Tuff.

mudstone. Rock made of silt and clay.

North American Plate. The lithospheric plate that includes most of North America and stretches roughly from the center of the Atlantic Ocean to the West Coast and from the Caribbean northward to the Arctic.

obsidian. A dark, glassy volcanic rock formed when liquid magma is quenched.

onlap. A geometric arrangement of sedimentary beds in which each successive bed extends farther horizontally than the bed below it before terminating, or pinching out. Onlapping beds commonly reflect deposition that occurred as sea level rose.

outcrop. Bedrock exposed at Earth's surface.

outlet glacier. A glacier issuing outward from a larger ice cap and flowing through a mountain pass or valley.

paleo. A prefix referring to the geologic past.

paleontologist. A scientist who studies fossil organisms.

pebble. A sedimentary particle with a diameter between 4 and 64 mm (about $3/16$ and $2\frac{1}{2}$ inches).

petrified. The state of organic matter that has been fossilized through conversion to stone. Petrification occurs when water containing dissolved inorganic matter saturates organic matter, such as wood, and replaces it, sometimes preserving internal structure such as tree rings and cells.

plagioclase. A very common feldspar mineral containing calcium and sodium.

plate. A relatively rigid segment of the lithosphere that moves independently of surrounding parts of the lithosphere; also called *lithospheric plate*.

plate tectonics. A widely accepted geologic theory in which the lithosphere is segmented into several large plates that glide horizontally on the underlying asthenosphere and move laterally relative to each other.

pluton. A body of igneous rock that cools and solidifies in the subsurface and consists of rock with a relatively uniform composition.

precipitate. A chemical reaction that produces a solid crystalline substance from a former liquid solution in which the ions making up the substance were dissolved.

pumice. A glassy volcanic rock containing so many bubble holes that it can float. Pumice forms when frothy magma, usually rhyolite, is quenched.

pyroclastic flow. A very hot mixture of volcanic ash, gas, and bits of magma that flows downslope during a volcanic eruption. Pyroclastic flows often form due to the collapse of an eruptive column.

quartz. The most common rock-forming mineral on Earth's surface. Quartz is composed of silicon and oxygen and typically is clear but also occurs in a wide variety of colors due to the inclusion of certain elements.

quench. To cool rapidly, often due to encountering water or ice.

radiocarbon dating. A method for dating organic material in which carbon-14 is used to determine the amount of time that has elapsed since the organism died.

recessional moraine. An accumulation of till that is left along the margins of a receding glacier during a pause in its retreat.

retreat (glacier). The shrinking that occurs along the margins of a glacier when loss of ice due to melting exceeds that formed through seasonal accumulation of snow.

reverse fault. A fault with a steep angle in which rocks situated above the fault are pushed upward relative to rocks below the fault. Reverse faults form due to tectonic compressional stress.

rhyolite. A fine-grained extrusive igneous rock rich in silicon and usually light colored. Compositionally, rhyolite is the equivalent of granite, which is an intrusive igneous rock.

ring fault. A steep fault that forms part of a roughly cylindrical network of faults that define the margin of a caldera and connect a shallow magma chamber with the ground surface.

rock. A naturally formed aggregate of one or more minerals.

sand. Sedimentary particles ranging between 0.06 and 2.5 mm in diameter (roughly $^2/_{1,000}$ and $^1/_{10}$ inch).

sandstone. A sedimentary rock dominated by sand-sized particles.

scarp. A long cliff or steep slope usually formed by offset of the ground surface by fault movement.

schist. A metamorphic rock characterized by well-defined, thin foliation and typically containing abundant crystals of mica that are oriented in the same direction.

sediment. Sedimentary particles that are eroded from preexisting rock and transported by wind, water, ice, or gravity; precipitated by chemical reactions; or secreted by organisms. Sediments accumulate as loose, unconsolidated layers on the Earth's surface.

sedimentary rock. Rock formed by the lithification of sediment.

seismic. Pertaining to earthquakes or vibrations caused by earthquakes.

shale. A fine-grained sedimentary rock composed of clay and silt.

silica. Silicon dioxide; a chemical compound that is made of silicon and oxygen and is the chief constituent of quartz, opal, and chert.

silicate. A class of minerals based on anions made of one silicon atom surrounded by four oxygen atoms.

siliceous sinter. See **sinter**

silt. Sedimentary particles ranging between 0.004 and 0.06 mm in diameter (about $^{15}/_{100,000}$ and $^2/_{1,000}$ inch).

sinter. Rock formed by the precipitation of silica in thermal water. *Siliceous sinter* is a synonym.

slump. A process in which a consolidated mass of rock and sediment moves downward, rotating along the scoop-shaped fracture along which it has been dislodged.

slump block. A coherent mass of sediment or rock displaced by slumping.

Snake River Plain. A topographic plain in southern Idaho that is about 400 miles (640 km) east-west and 75 miles (120 km) north-south and made mostly of basalt. It formed due to caldera-forming eruptions related to the Yellowstone hot spot and subsequent subsidence and basalt volcanism following the passage of the hot spot.

sorting. A process in which a moving fluid, such as water or wind, separates sediment according to size, density, or shape.

spreading center. A boundary where lithospheric plates are moving apart and new plate is forming. Spreading centers usually occur in the middle of ocean basins.

strata. Layers of sedimentary rock. **Beds** is a synonym.

subduction. The process by which a lithospheric plate descends into the asthenosphere beneath another lithospheric plate.

subduction zone. The region or boundary where one lithospheric plate descends beneath another.

supercontinent. A landmass composed of more than one continent.

supergroup. A formally defined, mappable body of rock ranking higher than and consisting of more than one group.

superheated. Water that is above the boiling point but does not boil because of confining pressure.

supersaturated. A chemical state in which the concentration of a particular dissolved ion exceeds the capacity of the fluid to keep the ion dissolved; usually leads to precipitation of the dissolved ion as crystals or amorphous substances such as sinter.

talus. Sloping accumulations of rock fragments at the base of a cliff or outcrop from which they have eroded.

tectonic. Pertaining to the forces involved in or the structures resulting from large-scale dynamic processes within the Earth; includes all lateral and vertical movements of the lithosphere (for example, mountain building). See also **plate tectonics**

terminal moraine. An accumulation of glacial till deposited at the end of a glacier; the moraine remains as a ridge or embankment after the glacier has retreated.

terrace. A relatively horizontal surface or level bench that typically breaks up an otherwise continuous slope. Terraces are formed by rivers, along lake or ocean shorelines, or by tectonic deformation.

texture. The physical characteristics of a rock, including grain or crystal size, shape, and arrangement.

thermal feature. A general term for any occurrence of thermal water at the Earth's surface, including hot springs, geysers, fumaroles, and mud pots.

thermal plume. A conduit or pipelike body of hot rock and gas moving upward from the mantle toward Earth's surface.

thermal water. Groundwater heated by or derived from magma near the ground surface.

thrust fault. A fault with a shallow angle along which rock moves horizontally and upward. Thrust faults form due to tectonic compressional stress.

till. Poorly sorted sediment of different sizes and types deposited directly by glacial ice.

travertine. A type of limestone (calcium carbonate), generally light colored, that precipitates from thermal water.

tuff. A general term for consolidated volcanic rock formed from pyroclastic flows. See also **ash-flow tuff**, **welded tuff**

unconformity. A break or gap in the geologic record, including that formed by an interruption in the deposition of sediments or a break between the erosion of older metamorphic or igneous rock and the deposition of younger sedimentary strata; an unconformity occurs as a physical contact (surface) separating the younger rock units above from the older ones below.

uplift. A structurally high part of the crust formed by tectonic movements. Also, the process by which rocks are raised to higher elevation.

valley glacier. A glacier that forms in mountainous terrain, generally originating in a cirque and flowing downslope into a stream valley.

vent. An opening at the Earth's surface from which volcanic materials or thermal water is extruded.

viscosity. The internal resistance of a substance to flow; the stiffness of a fluid.

volcanic arc. A chain of volcanoes on the margin of an overriding plate at a subduction zone. The magma supplied to the volcanoes comes from the melting of the plate that is being subducted.

volcanic ash. Solid particles, ranging from dust- to boulder-sized, that accumulate on the ground as fallout from an explosive volcanic eruption.

volcanic glass. Natural glass that lacks a crystalline structure. Glass forms when magma is quenched, cooling too rapidly for mineral crystals to form.

volcanic rock. Rocks formed by the extrusion of magma, solid debris (including volcanic ash), and hot gas from a volcano; includes intrusive rocks that cool at shallow depths and generally in association with a volcano.

weathering. The physical and chemical breakdown of rocks at Earth's surface due to contact with the atmosphere and the effects of freezing and thawing, wind, rain, chemical alteration, etc.

welded tuff. A dense, glassy volcanic rock formed from hot ash deposited by a pyroclastic flow. The heat of the deposit and weight of the overlying accumulation cause ash at lower stratigraphic levels to compact and partially fuse together.

Yellowstone Caldera. The now largely filled depression that formed when the Yellowstone Volcano collapsed following its eruption about 640,000 years ago.

Yellowstone Country. The region in and around Yellowstone National Park that includes the geology directly related to that which is exposed inside the park.

Yellowstone hot spot. The region at the Earth's surface characterized by elevated heat flow associated with the thermal plume under Yellowstone National Park.

Yellowstone hot spot track. The trail of extinct caldera-forming volcanoes that extends south and west of Yellowstone National Park down the length of the eastern Snake River Plain.

Yellowstone Plateau. The high-elevation region in central and southwestern Yellowstone National Park that is characterized by generally subdued topography and is mostly underlain by rhyolite lava flows of Pleistocene age. The plateau is roughly elliptical in shape: Its long axis stretches north and east from west of the southwestern Yellowstone National Park boundary for about 50 miles (80 km) to the Sour Creek resurgent dome; its short axis stretches south and east from the southern Gallatin Range for about 40 miles (64 km) roughly to the southeast shore of Yellowstone Lake.

Yellowstone Volcano. The large volcano that is roughly centered in Yellowstone National Park and is responsible for the massive caldera-forming eruptions and rhyolite flows in Yellowstone Country.

Sources of More Information

INTRODUCTION

Alt, D., and D. W. Hyndman. 1986. *Roadside Geology of Montana*. Roadside Geology Series. Missoula, Mont.: Mountain Press Publishing Co.

Christiansen, R. L. 2001. Geologic map of the Yellowstone Plateau area. In *The Quaternary and Pliocene Yellowstone Plateau Volcanic Field of Wyoming, Idaho, and Montana*, plate 1. U.S. Geological Survey Professional Paper 729-G.

Dickinson, W. R., and W. S. Snyder. 1978. Plate Tectonics of the Laramide orogeny. In *Laramide Folding Associated with Basement Block Faulting in the Western United States*, Geological Society of America Memoir 151, V. Matthews III, 355–66.

Fritz, W. J. 1985. *Roadside Geology of the Yellowstone Country*. Roadside Geology Series. Missoula, Mont.: Mountain Press Publishing Co.

Lageson, D.R., and D. R. Spearing. 1988. *Roadside Geology of Wyoming*. Roadside Geology Series. Missoula, Mont.: Mountain Press Publishing Co.

Locke, W. W., et al. 1995. The middle Yellowstone Valley from Livingston to Gardiner, Montana: a microcosm of northern Rocky Mountain geology. *Northwest Geology* 24:1–65.

Mallory, W. W., et al., eds. 1972. *Geologic Atlas of the Rocky Mountain Region*. Denver, Colo.: Rocky Mountain Association of Geologists.

Morgan, L. A., ed. 2007. *Integrated geoscience studies in the greater Yellowstone area—volcanic, tectonic, and hydrothermal processes in the Yellowstone geoecosystem*. U.S. Geological Survey Professional Paper 1717.

Reid, S. G., and D. J. Foote, eds. 1982. *Geology of the Yellowstone Park Area: Wyoming Geological Association 33rd Annual Field Conference Guidebook*. Wyoming Geological Association.

Smith, R. B, and L. J. Siegel. 2000. *Windows into the Earth: The Geologic Story of Yellowstone and Grand Teton National Parks*. USA: Oxford University Press.

U.S. Geological Survey. 1972. *Geologic Map of Yellowstone National Park*. Miscellaneous Geologic Investigations Map I-711.

1. GREAT UNCONFORMITY—SHOSHONE RIVER CANYON

Mueller, P. A., et al. 1998. Early Archean crust in the northern Wyoming province: Evidence from U–Pb ages of detrital zircons. *Precambrian Research* 91 (3–4):295–307.

Steward, J. H., and C. A. Suczek. 1977. Cambrian and latest Precambrian paleogeography and tectonics in the western United States. In *Paleozoic Paleogeography of the Western United States: Pacific Coast Paleogeography Symposium I*, eds. J. H. Stewart et al., 1–18. Los Angeles, Calif.: Society of Economic Paleontologists and Mineralogists, Pacific Section.

2. CAMBRIAN SEA LIFE—CODY, WY

Deland, C. R., and A. B. Shaw. 1956. Upper Cambrian trilobites from western Wyoming. *Journal of Paleontology* 30 (3):542–62.

Lochman, C., and C. Hu. 1962. Upper Cambrian faunas from the northwest Wind River Mountains, Wyoming, Part III. *Journal of Paleontology* 36 (1):1–28.

Miller, M. 1936. Cambrian trilobites from northwestern Wyoming. *Journal of Paleontology* 10 (1):23–34.

Sloss, L. L. 1963. Sequences in the cratonic interior of North America. *Geological Society of America Bulletin* 74 (2):93–114.

Thomas, R. C. 1995. Cambrian mass extinction ("biomere") boundaries: a summary of thirty years of research. *Northwest Geology* 24:67–75.

Vail, P. R., Mitchum, R. M., Jr., and S. Thompson, III. 1977. Global cycles of relative changes in sea level. In *Seismic Stratigraphy: Applications to Hydrocarbon Exploration*, American Association of Petroleum Geologists Memoir 26, ed. C. E. Payton, 83–98.

3. MISSISSIPPIAN LIMESTONE—PEBBLE CREEK CAMPGROUND

Peterson, J. A. 1977. Paleozoic shelf-margin and marginal basins, western Rocky Mountains-Great Basin, United States. In *Rocky Mountain Thrust Belt Geology and Resources: Wyoming Geological Association 29th Annual Field Conference in Conjunction with Montana Geological Society and Utah Geological Society*, eds. E. L. Heisey et al., 153–63. Casper: Wyoming Geological Society.

Poole, F. G., and C. A. Sandberg. 1977. Mississippian paleogeography and tectonics of the western United States. In *Paleozoic Paleogeography of the Western United States: Pacific Coast Paleogeography Symposium I*, eds. J. H. Stewart et al., 67–85. Los Angeles, Calif.: Society of Economic Paleontologists and Mineralogists, Pacific Section.

Rosen, P. R. 1977. Mississippian carbonate shelf margins, western United States. In *Rocky Mountain Thrust Belt Geology and Resources: Wyoming Geological Association 29th Annual Field Conference in Conjunction with Montana Geological Society and Utah Geological Society*, eds. E. L. Heisey et al., 155–72. Casper: Wyoming Geological Society.

4. CRETACEOUS INTERIOR SEAWAY—GARDNER RIVER CANYON

Lageson, D. R., and J. G. Schmitt. 1994. The Sevier orogenic belt of the western United States: recent advances in understanding its structural and sedimentologic framework. In *Mesozoic Systems of the Rocky Mountain Region, USA*, eds. M. V. Caputo et al., 27–64. Denver, Colo.: Society for Sedimentary Geology (SEPM), Rocky Mountain Section.

Lawton, T. F. 1994. Tectonic setting of Mesozoic sedimentary basins, Rocky Mountain region, United States. In *Mesozoic Systems of the Rocky Mountain Region, USA*, eds. M. V. Caputo et al., 1–25. Denver, Colo.: Society for Sedimentary Geology (SEPM), Rocky Mountain Section.

Peterson, J. A. 1994. Regional paleogeologic and paleogeographic maps of the Mesozoic systems, Rocky Mountain Region, U.S. In *Mesozoic Systems of the Rocky Mountain Region, USA*, eds. M. V. Caputo et al., 65–71. Denver, Colo.: Society for Sedimentary Geology (SEPM), Rocky Mountain Section.

5. SPHINX MOUNTAIN AND THE HELMET

DeCelles, P. G. 1986. Sedimentation in a tectonically partitioned, nonmarine foreland basin: the Lower Cretaceous Kootenai Formation, southwestern Montana. *Geological Society of America Bulletin* 97 (8):911–31.

DeCelles, P. G., et al. 1987. Laramide thrust-generated alluvial-fan sedimentation, Sphinx Conglomerate, Southwestern Montana. *American Association of Petroleum Geologists Bulletin* 71 (2):135–55.

Graham, S. A., et al. 1986. Lithology of source terranes as a determinant in styles of foreland sedimentation. In *Foreland Basins*, Special Publication of the International Association of Sedimentologists, no. 8, eds. P. A. Allen and P. Homewood. Blackwell Scientific Publications.

Ingersoll, R. V., et al. 1987. Provenance of impure calclithites in the Laramide foreland of southwestern Montana. *Journal of Sedimentary Petrology* 57 (6):995–1003.

6. RATTLESNAKE AND CEDAR MOUNTAINS

DeCelles, P. G., et al. 1991. Kinematic history of a foreland uplift from Paleocene synorogenic conglomerate, Beartooth Range, Wyoming and Montana. *Geological Society of America Bulletin* 103 (11):1458–75.

Erslev, E. A. 1986. Basement balancing of Rocky Mountain foreland uplifts. *Geology* 14 (3):259–62.

7. UNUSUAL VOLCANIC ROCKS NEAR SYLVAN PASS

Feeley, T. C. 2003. Origin and tectonic implications of across-strike geochemical variations in the Eocene Absaroka volcanic province, United States. *Journal of Geology* 111 (3):329–46.

Feeley, T. C., and M. A. Cosca. 2003. Time vs. composition trends of magmatism at Sunlight volcano, Absaroka volcanic province, Wyoming. *Geological Society of America Bulletin* 115 (6):714–28.

Haeussler, P. J., et al. 2003. Life and death of the Resurrection Plate: evidence for its existence and subduction in the northeastern Pacific in Paleocene-Eocene time. *Geological Society of America Bulletin* 115 (7):867–80.

Smedes, H. W., and H. J. Prostka. 1972. *Stratigraphic framework of the Absaroka Volcanic Supergroup in the Yellowstone National Park region.* U.S. Geological Survey Professional Paper 729-C.

8. DEBRIS FLOW DEPOSITS BETWEEN CODY AND EAST ENTRANCE

Fritz, W. J. 1980. Depositional environment of the Eocene Lamar River Formation in Yellowstone National Park. Unpublished PhD dissertation, University of Montana.

Fritz, W. J. 1980. Reinterpretation of the depositional environment of the Yellowstone "fossil forests." *Geology* 8 (7):309–13.

Fritz, W. J. 1982. Geology of the Lamar River Formation, Northeastern Yellowstone National Park. In *Geology of the Yellowstone Park Area: Wyoming Geological Association 33rd Annual Field Conference Guidebook*, eds. S. G. Reid and D. J. Foote, 73–101. Wyoming Geological Association.

Pierson, T. C., and K. M. Scott. 1985. Downstream dilution of a lahar: transition from debris flow to hyperconcentrated streamflow. *Water Resources Research* 21 (10):1511–24.

9. PETRIFIED WOOD IN TOM MINER BASIN

Chadwick, A., and T. Yamamoto. 1984. A paleoecological analysis of the petrified trees in the Specimen Creek area of Yellowstone National Park, Montana, U.S.A. *Palaeogeography, Palaeoclimatology, Palaeoecology* 45 (1):39–48.

Coffin, H. G. 1976. Orientation of trees in the Yellowstone petrified forests. *Journal of Paleontology* 50 (3):539–43.

Wikipedia contributors. "Paleocene-Eocene Thermal Maximum," *Wikipedia, The Free Encyclopedia*, http://en.wikipedia.org/wiki/Paleocene–Eocene_Thermal_Maximum (accessed 2010).

10. LANDSCAPE CHANGES AT HEPBURN'S MESA

Barnosky, A. D. 1982. Locomotion in moles (Insectivora, Proscalopidae) from the Middle Tertiary of North America. *Science* 216 (4542):183–85.

Barnosky, A. D., and W. J. Labar. 1989. Mid-Miocene (Barstovian) environmental and tectonic setting near Yellowstone Park, Wyoming and Montana. *Geological Society of America Bulletin* 101 (11):1448–56.

Burbank, D. W., and A. D. Barnosky. 1990. The magnetochronology of Barstovian mammals in southwestern Montana and implications for the initiation of Neogene crustal extension in the northern Rocky Mountains. *Geological Society of America Bulletin* 102 (8):1093–1104.

11. THE FIRST CALDERA-FORMING ERUPTION IN YELLOWSTONE

Bindeman, I. N., et al. 2007. Voluminous low $\delta18O$ magmas in the late Miocene Heise volcanic field, Idaho: implications for the fate of Yellowstone hotspot calderas. *Geology* 35 (11):1019–22.

Johnson, S. Y., et al. 2003. Hydrothermal and tectonic activity in northern Yellowstone Lake, Wyoming. *Geological Society of America Bulletin* 115 (8):954–71.

Pierce, K. L., et al. 2007. The Yellowstone hotspot, greater Yellowstone ecosystem, and human geography. In *Integrated geoscience studies in the greater Yellowstone area—volcanic, tectonic, and hydrothermal processes in the Yellowstone geoecosystem*, ed. L. A. Morgan, 1–39. U.S. Geological Survey Professional Paper 1717.

Sarna-Wojcicki, A. M., and J. O. Davis. 1991. Quaternary tephrochronology. In *Quaternary Nonglacial Geology: Conterminous U.S.*, The Geology of North America, vol. K-2, ed. R. P. Morrison, 93–116. Boulder, Colo.: Geological Society of America.

Sears, J. W. 2008. *Great elliptical basin, western United States: evidence for top-down control of the Yellowstone hot spot and Columbia River basalts.* www.MantlePlumes.org. http://www.mantleplumes.org/CRBEllipse.html.

12. THE YELLOWSTONE VOLCANO ERUPTS AGAIN

Christiansen, R. L. 2003. *The Quaternary and Pliocene Yellowstone Plateau Volcanic Field of Wyoming, Idaho, and Montana.* U.S. Geological Survey Professional Paper 729-G.

Smith, R. L., and R. A. Bailey. 1968. Resurgent Cauldrons. In *Studies in Volcanology, a Memoir in Honor of Howel Williams,* Geological Society of America Memoir 116, eds. R. R. Coats, R. R. Hay, and C. A. Anderson, 613–62.

13. RHYOLITE FLOWS IN THE FIREHOLE RIVER DRAINAGE

Dzurisin, D., Savage, J. C., and R. O. Fournier. 1990. Recent crustal subsidence at Yellowstone Caldera, Wyoming. *Bulletin of Volcanology* 52:247–70.

Gansecki, C. A., Mahood, G. A., and M. O. McWilliams. 1996. ^{40}Ar/^{39}Ar geochronology of rhyolites erupted following collapse of the Yellowstone Caldera, Yellowstone Plateau Volcanic Field: implications for crustal contamination. *Earth and Planetary Science Letters* 142 (1–2):91–107.

Pelton, J. R., and R. B. Smith. 1982. Contemporary vertical surface displacements in Yellowstone National Park. *Journal of Geophysical Research* 87 (B4):2745–61.

Waite, G. P., and R. B. Smith. 2002. Seismic evidence for fluid migration accompanying subsidence of the Yellowstone Caldera. *Journal of Geophysical Research* 107 (B9):2177.

Waite, G. P., Smith, R. B., and R. M. Allen. 2006. Vp and Vs structure of the Yellowstone hot spot from teleseismic tomography: evidence for an upper mantle plume. *Journal of Geophysical Research* 111 (B0):4303.

Wicks, C., Jr., Thatcher, W., and D. Dzurisin. 1998. Migration of fluids beneath Yellowstone Caldera inferred from satellite radar interferometry. *Science* 282:458–62.

Wicks, C. W., et al. 2006. Uplift, thermal unrest, and magma intrusion at Yellowstone Caldera. *Nature* 440:72–75.

14. ICE SCULPTURES ALONG THE BEARTOOTH HIGHWAY

Good, J. M., and K. Pierce. 1996. *Interpreting the Landscape: Recent and Ongoing Geology of Grand Teton and Yellowstone National Parks.* Moose, Wyo.: Grand Teton Natural History Association.

Pierce, K. L. 1979. *History and dynamics of glaciation in the northern Yellowstone National Park area.* U.S. Geological Survey Professional Paper 729 F.

Pierce, K. L. 2004. Pleistocene Glaciations of the Rocky Mountains. In *The Quaternary Period in the United States.* Developments in Quaternary Sciences Series, vol. 1, A. R. Gillespie, S. C. Porter, and B. F. Atwater, 63–76. New York: Elsevier.

15. GLACIAL DEPOSITS IN NORTHERN YELLOWSTONE

Licciardi, J. M., et al. 2001. Cosmogenic ^{3}He and ^{10}Be chronologies of the late Pinedale northern Yellowstone Ice Cap, Montana, USA. *Geology* 29 (12):1095–98.

Licciardi, J. M., and K. L. Pierce. 2008. Cosmogenic exposure-age chronologies of Pinedale and Bull Lake glaciations in greater Yellowstone and the Teton Range USA. *Quaternary Science Reviews* 27:814–31.

Richmond, G. M. 1964. Glacial geology of the West Yellowstone Basin and adjacent parts of Yellowstone National Park. In *The Hebgen Lake, Montana, earthquake of August 17, 1959,* 223–36. U.S. Geological Survey Professional Paper 435-T.

16. FLOODS AND EARTHFLOWS NEAR GARDINER, MT

Boulton, G. S., and R. C. A. Hindmarsh. 1987. Sediment deformation beneath glaciers: rheology and geological consequences. *Journal of Geophysical Research* 92 (B9):9059–82.

Hart, J. K. 1995. Subglacial erosion, deposition, and deformation associated with deformable beds. *Progress in Physical Geography* 19 (2):173–91.

Rudoy, A. N., and V. R. Baker. 1993. Sedimentary effects of cataclysmic late Pleistocene glacial outburst flooding, Altay Mountains, Siberia. *Sedimentary Geology* 85 (1–4):53–62.

Shaw, J. 2002. The meltwater hypothesis for subglacial bedforms. *Quaternary International* 90 (1):5–22.

17. MAMMOTH HOT SPRINGS

Andrews, J. E., and R. Riding. 2001. Depositional facies and aqueous-solid geochemistry of travertine-depositing hot springs (Angel Terrace, Mammoth Hot Springs, Yellowstone National Park, U.S.A.): discussion. *Journal of Sedimentary Research* 71 (3):496–97.

Chafetz, H. S., and R. L. Folk. 1984. Travertines: depositional morphology and the bacterially constructed constituents. *Journal of Sedimentary Research* 54 (1):289–316.

Chafetz, H. S., and S. A. Guidry. 1999. Bacterial shrubs, crystal shrubs, and ray-crystal shrubs: bacterial vs. abiotic precipitation. *Sedimentary Geology* 126 (1–4):57–74.

Chafetz, H. S., and S. A. Guidry. 2003. Deposition and diagenesis of Mammoth Hot Springs travertine, Yellowstone National Park, Wyoming, U.S.A. *Canadian Journal of Earth Sciences* 40 (11):1515–29.

Fouke, B. W., et al. 2000. Depositional facies and aqueous-solid geochemistry of travertine-depositing hot springs (Angel Terrace, Mammoth Hot Springs, Yellowstone National Park, U.S.A.) *Journal of Sedimentary Research* 70 (3):565–85.

Fouke, B. W. 2001. Depositional facies and aqueous-solid geochemistry of travertine-depositing hot springs (Angel Terrace, Mammoth Hot Springs, Yellowstone National Park, U.S.A.): reply. *Journal of Sedimentary Research* 71 (3):497–500.

Fouke, B. W., et al. 2003. Partitioning of bacterial communities between travertine depositional facies at Mammoth Hot Springs, Yellowstone National Park, U.S.A. *Canadian Journal of Earth Sciences* 40 (11):1531–48.

Husen, S., Smith, R. B., and G. P. Waite. 2004. Evidence for gas and magmatic sources beneath the Yellowstone Volcanic Field from seismic tomographic imaging. *Journal of Volcanology and Geothermal Research* 131 (3–4):397–410.

Jettestuen, E. et al. 2006. Growth and characterization of complex mineral surfaces. *Earth and Planetary Science Letters* 249 (1–2):108–18.

Sorey, M. L., ed. 1991. *Effects of potential geothermal development in the Corwin Springs known geothermal resources area, Montana, on the thermal features of Yellowstone National Park.* U.S. Geological Survey Water-Resources Investigations Report 91–4052.

18. SINTER—OLD FAITHFUL AND UPPER GEYSER BASIN

Braunstein, D., and D. R. Lowe. 2001. Relationship between spring and geyser activity and the deposition and morphology of high temperature (>73°C) siliceous sinter, Yellowstone National Park, Wyoming, U.S.A. *Journal of Sedimentary Research* 71 (5):747–63.

Guidry, S. A., and H. S. Chafetz. 2003. Siliceous shrubs in hot springs from Yellowstone National Park, Wyoming, U.S.A. *Canadian Journal of Earth Sciences* 40 (11):1571–83.

Hinman, N. W., and R. F. Lindstrom. 1996. Seasonal changes in silica deposition in hot springs systems. *Chemical Geology* 132 (1–4):237–46.

Hurwitz, S., et al. 2008. Climate-induced variations of geyser periodicity in Yellowstone National Park, USA. *Geology* 36 (6):451–54.

Husen, S., et al. 2004. Changes in geyser eruption behavior and remotely triggered seismicity in Yellowstone National Park produced by the 2002 M 7.9 Denali fault earthquake, Alaska. *Geology* 32 (6):537–40.

Hutchinson, R. A., Westphal, J. A., and S. W. Kieffer. 1997. In situ observations of Old Faithful Geyser. *Geology* 25 (10):875–78.

Jones, B., and R. W. Renaut. 2003. Hot spring and geyser sinters: the integrated product of precipitation, replacement, and deposition. *Canadian Journal of Earth Sciences* 40 (11):1549–69.

Lowe, D. R., and D. Braunstein. 2003. Microstructure of high-temperature (>73°C) siliceous sinter deposited around hot springs and geysers, Yellowstone National Park: the role of biological and abiological processes in sedimentation. *Canadian Journal of Earth Sciences* 40 (11):1611–42.

19. HYDROTHERMAL EXPLOSIONS—NORRIS GEYSER BASIN AND YELLOWSTONE LAKE'S NORTH SHORE

Browne, P. R. L., and J. V. Lawless. 2001. Characteristics of hydrothermal eruptions, with examples from New Zealand and elsewhere. *Earth-Science Reviews* 52 (4):299–331.

Fournier, R. O., et al. 1991. Conditions leading to a recent small hydrothermal explosion at Yellowstone National Park. *Geological Society of America Bulletin* 103 (8):1114–20.

Fournier, R. O., et al. 2002. *Results of weekly chemical and isotopic monitoring of selected springs in the Norris Geyser Basin, Yellowstone National Park during June-September, 1995.* U.S. Geological Survey Open-File Report 02–344. http://geopubs.wr.usgs.gov/open-file/of02-344/.

Friedman, I. 2007. Monitoring changes in geothermal activity at Norris Geyser Basin by satellite telemetry, Yellowstone National Park, Wyoming. In *Integrated geoscience studies in the greater Yellowstone area—volcanic, tectonic, and hydrothermal processes in the Yellowstone geoecosystem*, ed. L. A. Morgan, 514–32. U.S. Geological Survey Professional Paper 1717.

Kharaka, Y. K., Sorey, M. L., and J. J. Thordsen. Large-scale hydrothermal fluid discharges in the Norris-Mammoth corridor, Yellowstone National Park, USA. *Journal of Geochemical Exploration*, 69–70:201–5.

Marler, G. D. 1964. Effects of the Hebgen Lake earthquake of August 17, 1959, on the hot springs of the Firehole geyser basins, Yellowstone National Park. In *The Hebgen Lake, Montana, earthquake of August 17, 1959*, 185–97. U.S. Geological Survey Professional Paper 435-Q.

Morgan, L. A., et al. 2003. Exploration and discovery in Yellowstone Lake: results from high-resolution sonar imaging, seismic reflection profiling, and submersible studies. *Journal of Volcanology and Geothermal Research* 122 (3–4):221–42.

Morgan, L. A., et al. 2007. The floor of Yellowstone Lake is anything but quiet—new discoveries from high-resolution sonar imaging, seismic-reflection profiling, and submersible studies. In *Integrated geoscience studies in the greater Yellowstone area—volcanic, tectonic, and hydrothermal processes in the Yellowstone geoecosystem*, ed. L. A. Morgan, 91–126. U.S. Geological Survey Professional Paper 1717.

Morgan, L. A., Shanks, P. W. C., III, and K. L. Pierce. 2009. *Hydrothermal processes above the Yellowstone magma chamber: large hydrothermal systems and large hydrothermal explosions*, 1–95. Geological Society of America Special Paper 459.

Muffler, J. P., White, D. E., and A. H. Truesdell. 1971. Hydrothermal explosion craters in Yellowstone National Park. *Geological Society of America Bulletin* 82 (3):723–40.

Ohsawa, S., et al. 2002. Rayleigh scattering by aqueous colloidal silica as a cause for the blue color of hydrothermal water. *Journal of Volcanology and Geothermal Research* 113 (1–2):49–60.

Pierce, K. L., and L. A. Morgan. 1992. The track of the Yellowstone hot spot: volcanism, faulting, and uplift. In *Regional Geology of Eastern Idaho and Western Wyoming*, eds. P. K. Link, M. A. Kuntz, and L. B. Platt, 1–53. Geological Society of America Memoir 179.

White, D. E, Hutchinson, R. A., and T. E. C. Keith. 1988. *The geology and remarkable thermal activity of Norris Geyser Basin, Yellowstone National Park, Wyoming.* U.S. Geological Survey Professional Paper 1456.

20. THE 1959 HEBGEN LAKE EARTHQUAKE

Ball, V. 1960. *Montana Earthquake at Hebgen Lake: I Was There, My Personal Story.* published by Verla Ball, 23 pages.

Da Costa, J. A. 1964. Effect of Hebgen Lake earthquake on water levels in wells in the United States. In *The Hebgen Lake, Montana, earthquake of August 17, 1959*, 167–78. U.S. Geological Survey Professional Paper 435-O.

Fraser, G. D. 1964. Intensity, magnitude, and ground breakage. In *The Hebgen Lake, Montana, earthquake of August 17, 1959*, 31–35. U.S. Geological Survey Professional Paper 435-F.

Fraser, G. D., Witkind, I. J., and W. H. Nelson. 1964. A geological interpretation of the epicentral area—the dual-basin concept. In *The Hebgen Lake, Montana, earthquake of August 17, 1959*, 99–106. U.S. Geological Survey Professional Paper 435-J.

Hadley, J. B. 1964. Landslides and related phenomena accompanying the Hebgen Lake earthquake of August 17, 1959. In *The Hebgen Lake, Montana, earthquake of August 17, 1959*, 109–39. U.S. Geological Survey Professional Paper 435-K.

Hanly, T. F. 1964. Sediment studies on the Madison River after the Hebgen Lake earthquake. In *The Hebgen Lake, Montana, earthquake of August 17, 1959*, 151–65. U.S. Geological Survey Professional Paper 435-M.

Jackson, W. H. 1964. Changes in the floor of Hebgen Lake. In *The Hebgen Lake, Montana, earthquake of August 17, 1959*, 51–54. U.S. Geological Survey Professional Paper 435-H.

Murphy, L. M., and R. J. Brazee. 1964. Seismological investigations of the Hebgen Lake earthquake. In *The Hebgen Lake, Montana, earthquake of August 17, 1959*, 13–17. U.S. Geological Survey Professional Paper 435-C.

Myers, W. B., and W. Hamilton. 1964. Deformation accompanying the Hebgen Lake earthquake of August 17, 1959. In *The Hebgen Lake, Montana, earthquake of August 17, 1959*, 55–98. U.S. Geological Survey Professional Paper 435-I.

Ross, C. P., and W. H. Nelson. 1964. Regional seismicity and brief history of Montana earthquakes. In *The Hebgen Lake, Montana, earthquake of August 17, 1959*, 25–30. U.S. Geological Survey Professional Paper 435-E.

Ryall, A. 1962. The Hebgen Lake, Montana, earthquake of August 18, 1959: P waves. *Bulletin of the Seismological Society of America* 52 (2):235–71.

Stermitz, F. 1964. Effects of the Hebgen Lake earthquake on surface water. In *The Hebgen Lake, Montana, earthquake of August 17, 1959*, 139–50. U.S. Geological Survey Professional Paper 435-L.

Stewart, S. W., Hofmann, R. B., and W. H. Diment. 1964. Some aftershocks of the Hebgen Lake earthquake. In *The Hebgen Lake, Montana, earthquake of August 17, 1959*, 19–24. U.S. Geological Survey Professional Paper 435-D.

Tocher, D. 1962. The Hebgen Lake, Montana, earthquake of August 17, 1959, MST. *Bulletin of the Seismological Society of America* 52 (2):153–62.

U.S. Geological Survey. 1964. Map of Hebgen Lake–West Yellowstone area showing surficial features and ground deformation that resulted from the Hebgen Lake earthquake. In *The Hebgen Lake, Montana, earthquake of August 17, 1959*, plate 2. U.S. Geological Survey Professional Paper 435.

Witkind, I. J., et al. 1962. Geologic features of the earthquake at Hebgen Lake, Montana, August 17, 1959. *Bulletin of the Seismological Society of America* 52 (2):163–80.

Witkind, I. J. 1964. Events on the night of August 17, 1959—the human story. In *The Hebgen Lake, Montana, earthquake of August 17, 1959*, 1–4. U.S. Geological Survey Professional Paper 435-A.

Witkind, I. J. 1964. Reactivated faults north of Hebgen Lake. In *The Hebgen Lake, Montana, earthquake of August 17, 1959*, 37–50. U.S. Geological Survey Professional Paper 435-G.

Witkind, I. J. 1964. Structural damage in the Hebgen Lake–West Yellowstone area. In *The Hebgen Lake, Montana, earthquake of August 17, 1959*, 5–11. U.S. Geological Survey Professional Paper 435-B.

EPILOGUE

Lowenstern, J. B., et al. 2005. *Steam explosions, earthquakes, and volcanic eruptions—what's in Yellowstone's future?* U.S. Geological Survey Fact Sheet 2005–3024.

Millspaugh, S. H., Whitlock, C., and P. J. Bartlein. 2000. Variations in fire frequency and climate over the past 17,000 years in central Yellowstone National Park. *Geology* 28 (3):211–14.

National Park Service, Yellowstone National Park, 1988. *The Yellowstone Fires: a Primer on the 1988 Fire Season*, 18 pages.

Whitlock, C. 1993. Postglacial vegetation and climate of Grand Teton and southern Yellowstone National parks. *Ecological Monographs* 63 (2):173–98.

Whitlock, C., and P. J. Bartlein. 1993. Spatial variations of Holocene climate change in the Yellowstone region. *Quaternary Research* 39 (2):231–38.

FIGURE ACKNOWLEDGMENTS

The author wishes to acknowledge the published works below from which each of the indicated figures was modified for the purposes of this book. The full citation for each published work can be found under the listed vignette in the preceding pages.

PAGE	FIGURE DESCRIPTION	REFERENCE (and vignette number)
29	North American Cambrian and Ordovician sedimentary sequences	Sloss, 1963 (2)
38	North American sedimentary sequences	Sloss, 1963 (2)
48	history of global sea level fluctuations	Vail et al., 1977 (2)
61	thrust fault formation and conglomerate deposition	Graham et al., 1986 (5)
63	geologic map of Sphinx Mountain and The Helmet area	Graham et al., 1986 (5)
67	emplacement of Scarface and Shedhorn Mountain thrust sheets	Graham et al., 1986 (5)
73	steep vs. shallow subduction angle	Dickinson and Snyder, 1978 (Introduction)
76	Rattlesnake Mountain cross section	Erslev, 1986 (6)
82	simplified map of Absaroka Volcanic Supergroup	Feely and Cosca, 2003 (7)
87	simplified geologic map, Sylvan Pass area	U.S. Geological Survey, 1972 (Introduction)
92	cross section of volcanoes and alluvial apron	Smedes and Prostka, 1972 (7)
104	global temperature fluctuations	Wikipedia contributors, 2010 (9)
129	calderas of the Yellowstone Volcano	Smith and Siegel, 2000 (Introduction)
130	Yellowstone hot spot track	Pierce and Morgan, 1992 (19)
131	image of thermal plume	Waite et al., 2006 (13)
133	distribution of volcanic ash	Sarna-Wojcicki and Davis, 1991 (11)
137	thickness of Member A of HRT	Christiansen, 2001 (Introduction)
138	thickness of Member B of HRT	Christiansen, 2001 (Introduction)
146	cross section of crust under Yellowstone	Waite and Smith, 2002 (13)
150	thickness of Member A of LCT	Christiansen, 2001 (Introduction)
160	rhyolite magma chamber under Yellowstone	Husen et al., 2004 (17)
170	map of Quaternary rhyolite flows	Christiansen, 2001 (Introduction)
176	Yellowstone Ice Cap	Good and Pierce, 1996 (14)
228	cross section of Old Faithful's conduit	Hutchinson et al., 1997 (18)
229	pressure and temperature graphs	Hutchinson et al., 1997 (18)
254	distance fault blocks dropped during Hebgen Lake earthquake	U.S. Geological Survey, 1964 (20)
255	trenches formed along fault scarps	Myers and Hamilton, 1964 (20)
265	map and cross section of the Madison landslide	Hadley, 1964 (20)

GPS Coordinates for Stop Locations

STOP NUMBER	STOP DESCRIPTION	UTM NORTHING NAD 1983	UTM EASTING NAD 1983	LATITUDE deg. decimal-min.	LONGITUDE deg. decimal-min.
1. The Missing Record of Deep Time: The Great Unconformity in Shoshone River Canyon					
1	lower gate	4930016	645572	44 30.527 N	109 10.115 W
2	Great Unconformity	4930015	645524	44 30.527 N	109 10.151 W
3	cliffs of Precambrian rock	4929987	645281	44 30.515 N	109 10.335 W
2. Invasion of the Trilobites: Cambrian Sea Life Flourishes near Cody					
1	Gros Ventre Formation outcrop	4930019	645625	44 30.528 N	109 10.075 W
2	Gallatin Formation outcrop	4930273	646852	44 30.650 N	109 9.145 W
3. Limy Record of Shallow Seas: Mississippian Limestone at Pebble Creek Campground					
1	Madison limestone roadcut	4974130	570551	44 55.026 N	110 6.369 W
2	Madison limestone outcrop	4974178	570570	44 55.052 N	110 6.354 W
4. The Cretaceous Interior Seaway: Marine Deposits in Gardner River Canyon					
1	Eagle Nest Rock	4983497	524272	45 00.270 N	110 41.521 W
1	closer view, Eagle Nest Rock	4983698	524154	45 00.379 N	110 41.610 W
2	view of Mt. Everts	4983698	524154	44 58.106 N	110 42.455 W
5. Mountains Reduced to Rubble: Sphinx Mountain and The Helmet					
1	Johnny Ridge Road overlook	4993329	447385	45 5.488 N	111 40.118 W
2	Sphinx Mtn.—The Helmet saddle	5001360	460870	45 9.878 N	111 29.874 W
6. Basement Rock on the Rise: Uplift of Rattlesnake and Cedar Mountains					
1	view of Z-fold	4929177	644230	44 30.090 N	109 11.142 W
2	view of Rattlesnake Mountain Fault	4929148	644180	44 30.075 N	109 11.180 W
7. Strange Brew: The Unusual Volcanic Rocks near Sylvan Pass					
1	road to Lake Butte overlook	4928578	558077	44 30.491 N	110 16.161 W
2	volcanic breccia roadcut	4923253	570578	44 27.548 N	110 06.769 W
3	basalt roadcut	4924337	574170	44 28.112 N	110 04.051 W
8. Debris Flow Deposits: Coarse Conglomerate between Cody and East Entrance					
1	view of hoodoos	4924006	613817	44 27.624 N	109 34.156 W
2	andesite dike	4924291	611911	44 27.796 N	109 35.590 W
3	Chimney Rock	4922990	594291	44 27.244 N	109 48.891 W
9. Fossilized Forests: Petrified Wood in Tom Miner Basin					
1	outcrop along trail	4997896	494467	45 8.071 N	111 4.222 W
2	outcrop along trail	4998013	494341	45 8.134 N	111 4.318 W
10. From Playa Lakes to Rushing Rivers: Landscape Changes Recorded at Hepburn's Mesa					
1	view of Hepburn's Mesa from US 89	5016377	513298	45 18.046 N	110 49.823 W
2	close-up view of Hepburn's Mesa	5016217	513752	45 17.959 N	110 49.476 W

11. Arrival of the Hot Spot: The First Caldera-Forming Eruption in Yellowstone

1	Upper Terrace Drive overlook	4979487	523058	44 58.107 N	110 42.456 W
2	view from Bunsen Peak Trail	4975649	521814	44 56.036 N	110 43.413 W
3	Rustic Falls parking area	4975733	521746	44 56.082 N	110 43.464 W

12. The Yellowstone Volcano Erupts Again! Tuff Smothers the Region and a Caldera Forms

1	pullout south of Beryl Spring	4946732	520275	44 40.420 N	110 44.652 W
1	best viewing spot	4946739	520284	44 40.424 N	110 44.645 W
2	pullout above Gibbon Falls	4945205	519874	44 39.596 N	110 44.959 W
3	Gibbon Falls parking lot	4944524	518095	44 39.231 N	110 46.307 W
sidebar	Purple Mountain Trailhead	4943674	511443	44 38.780 N	110 51.342 W
sidebar	Purple Mountain summit	4945121	511233	44 39.562 N	110 51.499 W

13. The Youngest Eruptions: Rhyolite Flows in the Firehole River Drainage

1	Mallard Lake Dome view	4920944	525761	44 26.480 N	110 40.577 W
2	Biscuit Basin flow outcrop	4930309	513200	44 31.559 N	110 50.033 W
2	tightly folded flow bands	4930433	513119	44 31.626 N	110 50.094 W
3	Nez Perce Creek flow outcrop	4942072	510845	44 37.915 N	110 51.796 W

14. Ice Sculptures along the Beartooth Highway: Glaciers Carve Yellowstone's Landscapes

1	Beartooth Pass pullout	4981375	621356	44 58.527 N	109 27.654 W
2	Beartooth Butte	4977728	611119	44 56.658 N	109 35.490 W
3	Pilot Peak overlook	4976412	607387	44 55.982 N	109 38.344 W
4	till and striated bedrock exposure	4969723	606697	44 52.376 N	109 38.953 W

15. Rivers of Dirty Ice: Glacial Deposits in Northern Yellowstone

1	Pine Creek valley moraines	5038580	535837	45 29.990 N	110 32.478 W
2	view of Chico moraine	5023692	523811	45 21.980 N	110 41.757 W
3	ground moraine near Slough Creek	4973035	550199	44 54.538 N	110 21.845 W

16. Melting Ice and Sliding Shale: Floods and Earthflows near Gardiner

1	view of flood bar from US 89	4989784	518741	45 3.676 N	110 45.718 W
2	view of earthflow from US 89	4988829	519574	45 3.159 N	110 45.085 W
3	flood bar up close	4988969	518474	45 3.236 N	110 45.923 W
sidebar	Bunsen Peak Trailhead	4975453	521471	44 55.931 N	110 43.674 W
sidebar	Bunsen Peak summit	4975372	523195	44 55.884 N	110 42.363 W

17. Terraced Travertines: Mammoth's Famous Hot Springs

1	New Blue Spring	4979543	523067	44 58.137 N	110 42.449 W
1	stream under boardwalk	4979373	523085	44 58.045 N	110 42.436 W
1	Cupid and Canary springs	4979345	523194	44 58.030 N	110 42.353 W
2	Terrace Mountain landslide	4976800	522328	44 56.657 N	110 43.019 W

18. Siliceous Sinter: Old Faithful and Upper Geyser Basin

1	Old Faithful Geyser	4922988	513780	44 27.604 N	110 49.607 W
2	Pump Geyser	4923395	513563	44 27.824 N	110 49.770 W
3	Doublet Pool	4923462	513557	44 27.860 N	110 49.774 W

19. Hydrothermal Explosions: Norris Geyser Basin and Yellowstone Lake's North Shore

1	Pearl Geyser	4952140	523171	44 43.336 N	110 42.445 W
2	Porkchop Geyser	4952151	523113	44 43.342 N	110 42.489 W
3	Indian Pond	4934275	553798	44 33.588 N	110 19.355 W
3	explosion breccia exposure	4933861	567173	44 33.297 N	110 19.254 W
4	Steamboat Point overlook	4930967	555836	44 31.792 N	110 17.837 W

20. The Night the Ground Shook: The 1959 Hebgen Lake Earthquake

1	Cabin Creek Fault Scarp Area	4968764	472963	44 52.307 N	111 20.537 W
2	Hebgen Dam	4967955	473594	44 51.871 N	111 20.055 W
3	Earthquake Lake Visitor Center	4964245	466265	44 49.849 N	111 25.606 W

Index

Page numbers in *italics* indicate an illustration or information in a caption.

About the Author

Marc S. Hendrix is a geology professor at The University of Montana in Missoula. Growing up in Gettysburg, Pennsylvania, Marc developed an early love of geology in the 1970s while working as a field assistant for his father, a biology professor at Gettysburg College. Marc received a bachelor's degree in geology from Wittenberg University in Springfield, Ohio, in 1985 and a master's degree in geology and geophysics from the University of Wisconsin in 1987. In 1992, he graduated with a PhD in applied earth sciences from Stanford University, where he conducted research on the geologic record of mountain building and ancient climate in central Asia, particularly in western China and Mongolia. In 1994, Marc joined the faculty at The University of Montana, where he has developed a field-based research program focused on the geology of the northern Rocky Mountains, including the Yellowstone region. Marc's work as a geologist has taken him to many remote corners of the world, particularly in Asia. He currently lives in Missoula with his wife and two sons.

—Matthew P. McArdle photo